# Advances in Turbomachinery

*Edited by Melih Cemal Kushan and Isil Yazar*

Published in London, United Kingdom

Advances in Turbomachinery
http://dx.doi.org/10.5772/intechopen.102305
Edited by Melih Cemal Kushan and Isil Yazar

Contributors
Edgar Jeevan Danaraj, Dmitry Alexandrovich Balakin, Alexey Zubko, Vitaly Shtykov, Shalimova Elena Vladimirovna, Zayed Saleh Salem Ali, Hernan Bolaños, Francisco Botero, Quanyong Xu, Honghu Wu, Anping Hou, Wang Di, Yang Zhen



First published in London, United Kingdom, 2023 by IntechOpen
IntechOpen is the global imprint of INTECHOPEN LIMITED, registered in England and Wales, registration number: 11086078, 5 Princes Gate Court, London, SW7 2QJ, United Kingdom

British Library Cataloguing-in-Publication Data
A catalogue record for this book is available from the British Library

Additional hard and PDF copies can be obtained from orders@intechopen.com

Advances in Turbomachinery
Edited by Melih Cemal Kushan and Isil Yazar
p. cm.
Print ISBN 978-1-83768-441-0
Online ISBN 978-1-83768-442-7
eBook (PDF) ISBN 978-1-83768-443-4

# Meet the editors

Melih Cemal Kushan is a Professor in the Department of Aeronautical Engineering, Eskisehir Osmangazi University, Turkey. He obtained a BSc in Mechanical Engineering from Dokuz Eylul University, Turkey, an MSc in Industrial Engineering from Anadolu University, Turkey, and a Ph.D. in Mechanical Engineering from Eskisehir Osmangazi University. He conducted his postdoctoral research at the College of Science and Engineering, Department of Mechanical Engineering, James Cook University, Australia. Dr. Kushan has more than 216 publications on aviation and defense technology to his credit. He has completed twenty-one research projects in materials science and the manufacturing of aerospace components.

Isil Yazar is an Associate Professor of Aeronautical Engineering at Eskisehir Osmangazi University, Turkey. She obtained a BS and MSc in Electrical and Electronics Engineering from the same university in 2005 and 2008, respectively. She also received an MBA from Anadolu University, Turkey, in 2008. From 2008 to 2010 she worked as a manufacturing engineer at Turkish Aerospace Industries (TUSAS). She was a member of AIRBUS A400M, ANKA UAV, and T-38 training aircraft projects. Dr. Yazar obtained a Ph.D. in Avionics from Anadolu University as a project researcher under the guarantee of the TEI/TUSAS-Anadolu University collaborative protocol in 2015. Her research interests include novel propulsion technologies in aviation, gas turbines, avionics, and unmanned aerial vehicles (UAVs).

# Contents

# Preface

Turbomachinery describes machines that transfer energy between a rotor and a fluid. Some of these machines include turbines, compressors, and pumps. Turbomachinery is important for industry and is critical in many different types of systems. This book discusses some turbomachinery problems and solutions. It is organized into two sections.

Section 1, "Turbomachinery Problems" includes three chapters: "Solid Particle Erosion", "Periodic Instabilities in a Specific Low-speed Pump Working as a Turbine", and "The Influences of Combined Distortions on Fan Performance". Section 2, "Turbomachinery Solutions" consists of two chapters: "Electrification for Aero-Engines: A Case Study of Modularization in New Product Development" and "Principles of Diagnosing: The Technical Condition of the Bearings of the Gas Turbine Engine Supports Using Rhythmogram and Scatterogram".

*Advances in Turbomachinery* is a current and well-rounded resource for students and practicing engineers in mechanical and aeronautical engineering as well as readers interested in turbomachinery problems and solutions.

**Melih Cemal Kushan**
Professor,
Department of Aeronautical Engineering,
Eskisehir Osmangazi University,
Eskisehir, Turkey

**Isil Yazar**
Associate Professor,
Department of Aeronautical Engineering,
Eskisehir Osmangazi University,
Eskisehir, Turkey

Section 1

# Turbomachinery Problems

Chapter 1

# Solid Particle Erosion

*Wang Di and Yang Zhen*

## Abstract

When military helicopters and transport aircraft take off and land forcibly on desert or simple runway, sand and dust will cause solid particle erosion damage to aircraft blades, leading to compressor performance degradation and structural integrity damage, which seriously affects the reliability of the engine. For the field of erosion protection, this book introduces the following six aspects, including erosion mechanism, influencing factors, protection methods, preparation methods, coating systems and structures, and the application status of erosion resistant coatings. The research and development direction of erosion resistant coating technology in the future was clarified, which laid a foundation for further research and application of erosion resistant coating technology.

**Keywords:** solid particle erosion, erosion mechanism, erosion resistant coating, erosion protection, coating systems and structures

## 1. Introduction

Solid particle erosion (SPE) is one of the common wear modes in engineering field. When solid particulate matter (sand, fly ash, salt, ice crystal, volcanic ash, etc., as shown in **Figure 1**) is entrained by air flow and impacts the surface of components, solid particle erosion will occur. In various applications, including helicopter rotor blades, wind turbines, power generation gas turbines, aircraft windshields, fuselage and engines, etc., they may be subject to severe erosion wear, resulting in the removal of component materials. As early as the Vietnam War in the 1960s, the US military had realized the seriousness of sand erosion. Because Vietnam is in a sand and dust environment, the compressor blades of T53 turboshaft engine of the "Huey" helicopter and the "Cobra" helicopter of the US military mission suffered sand and dust erosion, and serious geometric deformation and structural damage occurred, resulting in engine power reduction, blade crack failure, etc., which necessitated the engine to be replaced in advance, and the average maintenance interval was significantly shortened. By the 1970s, solid particle erosion had become an urgent problem in the field of aerospace. During the Gulf War, the T-64 engine of the US CH53E helicopter had serious erosion and wear problems of compressor blades in this desert environment, which reduced the operating time of the engine from 2000 hours in the ordinary environment to 100 hours in the desert environment. During the war in Afghanistan, the mission of the Russian Mi-17 helicopter in the sand and dust environment, a number of blade fracture failures occurred, resulting in a significant reduction in the service life of the engine and a serious threat to flight safety. In 2010, the Eyjafjallajokull volcano eruption in southern

**Figure 1.**
*Gravel, volcanic ash particles, salt and ice crystals erode engine parts to form SPE.*

Iceland led to the closure of the largest civil aviation line since the Second World War, which is precisely due to the reduced visibility of the region and the volcanic ash caused by the volcanic eruption easily entering the engine and endangering flight safety. It is generally believed that when the speed reaches 300 m/s, sand will inevitably enter the turbine engine [1]. Once the engine inhales sand dust, the sand will start to impact and slowly erode the blades. At the same time, corrosive liquids such as rain will accelerate its corrosion. On the contrary, erosion will accelerate the destruction of the integrity of the blades, and the corrosion will intensify. Thus, erosion and corrosion will act together to erode and destroy the blades in a domino manner, leading to catastrophic consequences of the engine.

In other fields, gas turbines are eroded by 5 μm coal fly ash particles. When particles are driven into the gas turbine by the gas flow, they will cause erosion damage to the static and moving blades, which will change the shape and size of the blades, leading to lower working efficiency and worse performance. In 1990, BP conducted an investigation on throttle valve failures in Alaska oil field, which showed that 34% of failures were caused by wear of valve internals and valve bodies [2]. However, in the pipeline system of pumps and heat exchangers, there are huge hidden dangers of safety accidents [3–6]. In a word, erosion wear is very harmful in the industrial production field [7]. In order to effectively reduce the loss caused by erosion wear and improve the service life and reliability of equipment and materials, scholars at home and abroad have carried out a variety of researches. Among them, hard coatings are increasingly used to improve the service life of components, which can significantly improve the anti-erosion performance by endowing the components with surface mechanical properties other than their own.

## 2. Brief introduction of SPE

### 2.1 Erosion mechanism

There have been a lot of studies on the solid particle erosion mechanism. Researchers found that under the action of sand and dust in complex environment, multiple erosion failure mechanisms are coupled, and a single erosion theory cannot meet the requirements of revealing the erosion phenomenon. At present, there are mainly the following erosion wear theories.

In 1985, the micro cutting theory came out. Subsequently, the erosion theory, secondary erosion theory, adiabatic shear and local deformation wear theory of plastic/brittle materials have developed successively. From the current erosion theory, micro cutting theory [8, 9] (as shown in **Figure 2a**) has analyzed the cutting action in

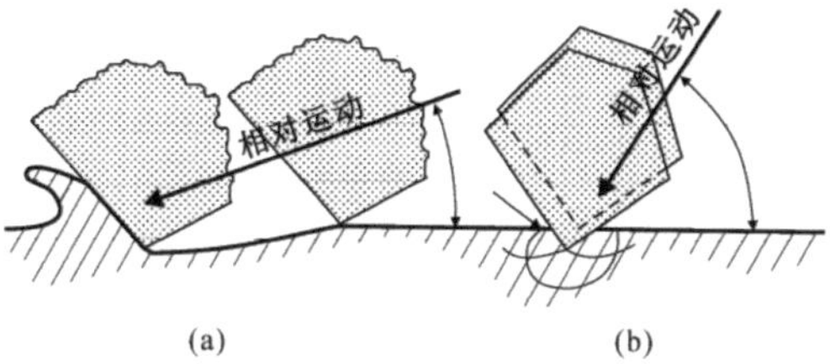


**Figure 2.**
*Erosion mechanism (a) low angle, and (b) high angle.*

the erosion process under the condition of small angle of attack (small angle), but there is a large error in explaining the erosion behavior at large angle of attack (large angle). The deformation and wear theory [10] revealed the deformation process and energy change analysis of samples after multiple impact angle erosion. The theoretical model believed that the micro cutting effect of particles was not obvious during large angle erosion, and the wear of materials was mainly related to the deformation caused by impact, which made the materials crack and fall off. The elastoplastic indentation fracture theory [11] analyzed the erosion of brittle materials by sand dust under the condition of high angle of attack (**Figure 2b**). The secondary erosion theory [12] provides a good explanation for the complex process of erosion process, namely, brittle particle fragmentation and re erosion of materials.

## 2.2 Factors affecting of SPE

The factors affecting the erosion behavior of materials are very complex, including external factors such as size and shape of erosion particles, as well as internal factors such as physical properties and microstructure of materials themselves. In addition, the parameters of erosion test such as erosion speed, angle, time, ambient temperature and other test factors will also affect the erosion results.

### *2.2.1 Erosion particle size and shape*

The influence of erosion particle size on materials with different properties is different. For ductile materials, the "size effect" of erosion particles [13], determines the erosion rate of the material, that is, the particle size increases, the erosion rate increases, but the critical size no longer meets the "size effect". The secondary erosion theory also believes that when the brittle particles reach a certain scale, they are easy to be broken in the erosion process. After the particles are broken, the erosion energy decreases, which is not enough to cause secondary erosion of materials, and the erosion rate will tend to be stable. For brittle materials, the erosion rate of small angle gradually increases with the decrease of particle size; Under the condition of large angle erosion, the erosion wear caused by large particles is more serious, which is usually a typical brittle fracture erosion mechanism. In addition, under the condition of the same particle size, the impact of particle shape on the material erosion rate is also not the same. Levy et al. [14], studied the influence of $Al_2O_3$ particles with different shapes on the erosion wear behavior of carbon steel. The results showed that the weight loss of irregular angular $Al_2O_3$ particles was much larger than that of spherical $Al_2O_3$ particles. Levy et al. believed that when angular $Al_2O_3$ particles impact the material surface, the contact area between the spherical particles and the matrix is much smaller, so the contact stress on the material surface is much larger under the same erosion conditions. The high stress concentration on the surface of the material makes it easier to initiate cracks, and then crack, resulting in a large amount of wear [11].

### 2.2.2 Impact of erosion parameters

When evaluating the erosion resistance of materials, erosion parameters (erosion angle, velocity, time, etc.) are the key factors affecting the erosion process of materials. For example, Finnie et al. [15], studied the erosion behavior of two representative materials, Al (ductile material) and $Al_2O_3$ (brittle material). It can be seen from **Figure 3** that the erosion rate of ductile material (Al) first increases with the increase of erosion angle and reaches the maximum value at about 15°, and then decreases with the increase of erosion angle. For brittle materials ($Al_2O_3$), the erosion rate increases with the increase of erosion angle. The results show that the erosion rate of brittle materials is lower at small angle of attack, while that of ductile materials is lower at large angle of attack. The analysis shows that the plowing effect of ductile materials with low surface hardness is significant in small angle erosion. Under the condition of large angle, the ductile material has a better inhibition effect on the generation of internal fatigue cracks, and the erosion rate is significantly reduced. On the contrary, the high hardness of brittle materials can resist the plowing effect of erosion at small angles of attack, but the high brittleness makes them vulnerable to impact fracture at large angles of attack, resulting in a large number of fatigue cracks and failure. Therefore, ductile materials and brittle materials are two distinct erosion behaviors under the same conditions.

In addition, particle velocity and erosion time are also the key factors affecting the erosion rate of materials. The general rule shows that the erosion rate increases gradually with the extension of erosion time. However, in the early stage of erosion, namely the so-called incubation period, the material will not suffer mass loss, and even a small amount of weight gain may occur. After initial erosion inoculation, the material enters a stable mass loss state. The length of incubation period is affected by erosion angle, erosion rate and material properties. The incubation period will be prolonged with the increase of erosion attack angle and the decrease of particle velocity. In addition, plastic materials usually have a longer incubation period than brittle materials [17]. For erosion particle velocity, lower velocity means lower impact energy, which makes it difficult to introduce high impact stress and reach the stress threshold of plastic deformation or crack initiation. Higher speed can cause surface

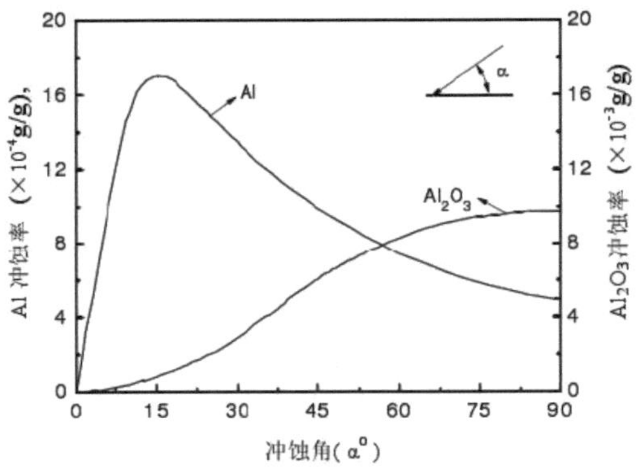


**Figure 3.**
*Relationship between erosion rate and angle [16].*

damage of materials. Therefore, there is a critical velocity for erosion particles. Below this speed, the particles will be bounced away without damage. Only after exceeding this speed, the material will start to suffer from erosion wear [16].

### 2.3 Methods of SPE protection

In order to enhance the erosion resistance of materials and prolong their service life, the erosion resistance can be improved by changing the overall structure of materials, but it is difficult to achieve for parts with more complex structures. As the erosion behavior first and mainly acts on the material surface, the surface quality of the material is the key to affect its erosion resistance. Adding protective coatings on the material surface to improve the erosion resistance is easier to achieve, and has been widely used. Surface coating technology is to obtain the required surface morphology, material composition, microstructure, etc. of parts and components through surface pretreatment, coating and other means. Therefore, it is of great significance to adopt modern surface modification technology for surface protection of related components.

## 3. Preparation method of SPE coating

The preparation method of the coating is physical vapor deposition (PVD) based on plasma technology. Physical vapor deposition is a common technology for preparing hard coatings. Its principles include: (1) gasification of coating materials, that is, providing energy to coating materials through collision, sputtering and other methods to ionize the coating materials and become a source of plasma gasification; (2) After sputtering and collision, the atoms or ions in the gasified coating materials are effectively separated from the plasma coating materials by the migration of atoms, molecules or ions in the coating materials, and the atoms or molecules in the coating materials are controlled to migrate to the substrate surface by magnetic field and other means; and (3) Coating atoms, molecules or ions are deposited on the body, that is, the separated atoms, molecules or ions are deposited or bombarded on the target surface by voltage or other methods. Physical vapor deposition (PVD) has become a promising coating preparation technology due to its high bonding strength, low deposition temperature, rich materials and multilayer coating.

The commonly used PVD technologies are mainly magnetron sputtering ion plating and arc ion plating. Magnetron sputtering ion plating is not suitable for preparing thick coatings because its ionization rate is much lower than that of arc ion plating due to its collision miss mechanism; At the same time, the toughness of the coating prepared by magnetron sputtering is far less than that of arc ion plating, and its erosion resistance is poor, so it is not suitable for the preparation of erosion resistant coatings. The other commonly used PVD technology, arc ion plating, is widely used for the preparation of hard coatings. Because the traditional cathodic arc ion plating is a thermal field emission miss target mechanism, the deposition particle energy is high, the flux is large, and the ionization rate is high (about 80%). At the same time, the technology can install different targets to achieve the deposition of multiple coatings, but the target will partially melt and eject micron sized molten metal particles due to arc discharge (as shown in **Figure 4**). These protrusions formed by micro-particles are easy to be eroded by sand and gravel, and will form pits after the micro-particles on the surface are washed away. These pits are easy to become stress concentration points of fatigue or erosion damage, thus leading to rapid failure of the coating.

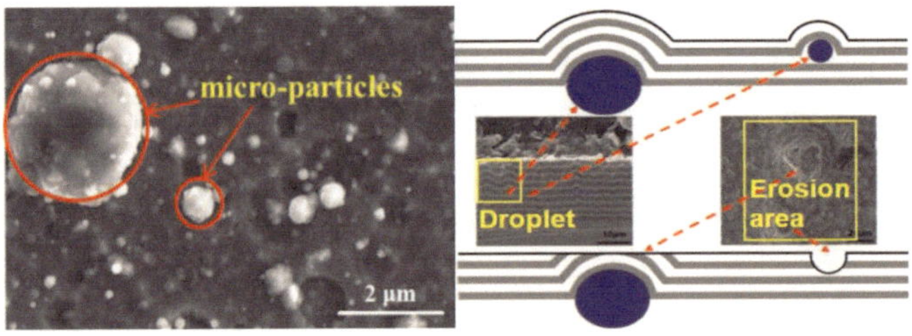


**Figure 4.**
*Schematic diagram of the surface morphology of the arc ion plating coating and the influence of large particles on the erosion.*

Therefore, how to eliminate or reduce the metal droplets in the anti-erosion coatings prepared by arc ion plating is one of the current research focuses.

## 4. Material system and structure design of SPE coating

### 4.1 Material system of erosion resistant coating

In the past 25 years, there have been many studies on anti-erosion coating material systems, most of which are comparative studies aimed at evaluating the performance of specific materials. Most studies focus on two coating series: one is carbon based system [2, 18, 19] (diamond, diamond-like carbon (DLC), tetrahedral amorphous carbon (ta-C) and some carbides); The second is a system based on nitrides (mainly TiN) [20–26] (TiN, TiAlN, TiCN, TiSiN, TiSiCN, etc.). Other nitrides (CrN [27, 28], ZrN [21], etc.) have also been partially studied.

Diamond coating has many excellent properties, most notably high hardness and strength, which makes it an attractive choice for friction and wear resistant parts. Therefore, many researchers have carried out detailed experimental studies on the erosion resistance of diamond coatings, and have a more in-depth understanding and analysis of the impact of erosion grit material, shape, size, erosion speed and angle on the erosion resistance and erosion mechanism.

Wheeler et al. [2] have studied the erosion resistance of diamond coatings in detail. They used Chemical vapor deposition technology to prepare diamond coatings of 10–46 μm thickness, and carried out erosion tests at 90° angles of attack using grit with different average diameters and velocities, the erosion resistance was studied. The erosion rate is related to the kinetic energy of the erosion particles, and compared with tungsten carbide and stainless steel. It is concluded that the erosion mechanism of diamond coating is composed of three stages. First, the coating produces microcracks; Secondly, pinholes and interfacial debonding occur; The last is the complete failure of the coating. Wheeler and others also studied the impact of different impact angles on the erosion performance of diamond coatings. The results show that although the number of impacts required at the beginning of pinhole increases significantly at small angles, all angles will produce a "pinhole" damage feature. As shown in **Figure 5**, circular cracks and pinholes are observed on the eroded surface, and can be observed on the eroded coating at all angles. As a part of establishing their formation mechanism, the author measured the diameter of annular cracks, analyzed the Hertz fracture theory and proposed the stress wave reflection theory. According to these results, the author also proposed to use energy to describe the impact of solid particles on the erosion resistance

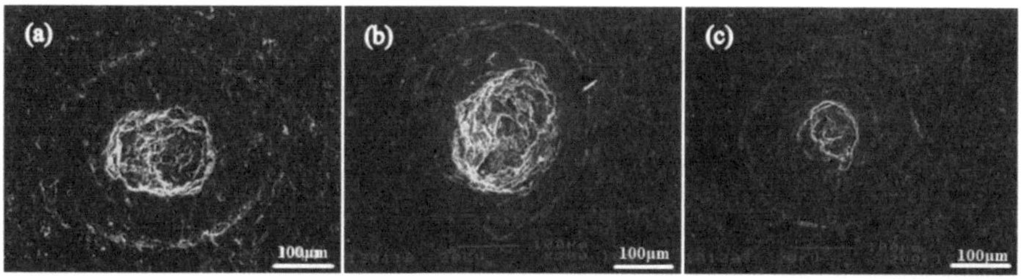


**Figure 5.**
*Micrographs of the surrounding cracks and pinholes of diamond coating under multi-angle impact [2] (a) 30°, (b) 60°, and (c) 90°.*

of the coating: $E_{EP} = E_c + E_E$, where $E_{EP}$ is considered to be the total energy consumed by the elastoplastic damage caused by a single impact, while $E_c$ and $E_E$ are the energy dissipated by the coating and erosion damage, respectively.

Till now, TiN based coatings have been the most widely used SPE resistant coatings in aeroengines. Therefore, many researchers still focus on TiN and its multiple systems. In the past 20 years, more and more researchers have begun to study nanocomposite TiN based coatings with better mechanical properties. The research results of Reed et al. [29] show that the erosion resistance of (Ti, Cr) N nano coating is highly dependent on the coating thickness and Cr content, and the increase of CrN phase volume reduces the hardness of the coating. Under the attack angle of 30°, all the coated samples are better than the uncoated substrate; However, under the condition of 90° attack angle, the coating deposited at low bias voltage (−25 V, −50 V, and −100 V) and Ti: Cr ratio (N = 2.4) is superior to the uncoated substrate. Therefore, hardness does not play a decisive role in the erosion resistance of the coating.

In recent years, CrN coating has attracted extensive attention of researchers because of its excellent toughness, wear resistance, high temperature oxidation resistance and corrosion resistance, and low residual stress of the coating, which makes it easy to deposit thicker coatings [30, 31]. However, the hardness of CrN coating is low, only about 1800–2000 Hv. At present, multiple nitrides [32–34] can be formed by doping elements (such as Al, Ti, TiAl, Si, etc.) to improve the hardness of the coating. Among them, Al and N in CrAlN coating are bound by covalent bond, and the grains of the coating are uniform and fine, which not only improves the hardness to about 2500–3000 Hv, but also increases the thermal stability of the coating. In the high temperature environment, Al atoms and Cr atoms are easy to diffuse outward, and combine with oxygen to form a more compact $Cr_2O_3$ and $Al_2O_3$ oxide layer. After the formation of this oxide layer, the volume expands, forming a compressive stress on the coating, which can resist the crack initiation on the surface, effectively prevent the deep level of oxidation, and improve the high-temperature oxidation resistance of the coating [35, 36]. According to the experience in preparing coatings in the field of engineering machinery, the most common method to improve the mechanical properties of metal nitrides, including hardness, toughness and oxidation resistance, is to add Al. Many studies show that the nitride coating containing Al still shows good wear resistance at high temperature [37–39]. Ren et al. [40] deposited CrAlN and CrN coatings on the steel substrate respectively. They found that the addition of Al makes some Cr atoms replaced by Al atoms to form CrAlN phase, which is conducive to refining grains and improving the comprehensive performance of CrAlN coatings. Wu et al. [41] deposited four kinds of coatings: TiN, TiAlN, CrAlN and CrAlTiN on the steel substrate. It is found that the CrAlN coating has the best erosion resistance, which is 3.9 times of the substrate.

According to the erosion model proposed by Evans et al. [42] and improved by Hockey et al. [43], they believe that the volume loss of brittle materials during erosion is proportional to the velocity, radius, density and impact angle of erosion particles, and inversely proportional to the fracture toughness and hardness of coating materials. Under the condition of high angle of attack, micro brittle fracture caused by elastic-plastic deformation is one of the main modes of erosion damage of hard coatings. According to the micro brittle fracture theory [42], the erosion volume (W) of materials can be expressed as:

$$W = CK^{-\frac{4}{3}}H^{-\frac{1}{4}} \tag{1}$$

Where: $W$ – Erosion volume; $H$ – hardness; $K$ – toughness; $C$ – Constant (depending on particle size, velocity and particle density of erosive particles). It can be seen from Eq. (1) that the erosion resistance of the coating can be improved by improving the hardness and toughness of the coating material. Moreover, improving the toughness has a more significant effect on improving the erosion resistance of the coating. Therefore, when selecting coating materials, it is necessary to comprehensively consider the hardness and toughness. On the basis of ensuring a certain hardness, it is necessary to focus on improving the toughness. It should be emphasized that it is meaningless to simply discuss toughness instead of strength (or hardness). Only those hard and tough coating materials have engineering application value.

### 4.2 Structural design of erosion resistant coating

With the deepening of research, some defects of monolayer hard coating in the field of erosion resistance have been revealed: the high stress of the coating, the high brittleness, and the low toughness, which leads to cracks easily appear when the coating is eroded by solid particles [44]. According to the literature, the maximum thickness of metal nitride coating with monolayer structure is about 6–8 μm [45]. However, when the coating thickness is thin, its erosion resistance is difficult to meet the protection requirements of related parts. At present, gradient multilayer coatings with complex structure have become the research focus of erosion resistant coatings. This kind of coating has a large number of interfaces, through which the continuous growth of columnar crystals can be restrained, the energy of erosion particles can be dissipated, and the initiation of crack sources and the propagation of buffer cracks can be prevented; At the same time, the interlayer can release the residual stress to a certain extent, coordinate the deformation, and improve the film substrate bonding strength and coating toughness.

According to the principle of multilayer strengthening, the difference of crystal structure and elastic modulus is the main reason for the dislocation to be blocked at the interface, thus forming the strengthening. In the multi-layer structure with alternating soft and hard, even if the metal (Me) in the metal layer and ceramic layer is the same element, the crystal structures of Me and MeN are different, so they will not grow epitaxial structures. The difference of elastic modulus will form a certain strengthening effect. Zhang et al. [46] studied the evolution of the cyclic impact damage mechanism of TiN/Ti multilayers with the sharp shortening of the modulation period from micrometer (1000 nm) to nanometer (60 nm). The results show that with the decrease of modulation period, the ductile phase of the films decreases, and the microstructure changes from TiN/TixNy/Ti to TiN/TixNy ($x > y$). The results of cyclic impact show that the impact resistance and damage mechanism of TiN/Ti multilayers are closely related to the modulation period. The smaller the modulation period of TiN/Ti multilayers, the lower the critical fracture load, the higher the fracture probability, and the worse the impact resistance.

Wieciński et al. [47] studied the erosion resistance and fracture mechanism of nanostructured Cr/CrN multilayer coatings. The researcher deposited seven multilayer coatings with different modulation ratios (Cr/CrN) by arc ion plating. These coatings have the same thickness (5–6 μm) and 16 layers of Cr/CrN, but the thickness ratio ($Q_{Cr/CrN}$) of Cr and CrN component layers is different. The small particles of silicon dioxide used in the erosion experiment impact the coating surface at an angle of attack of 90°. **Figure 6** shows the morphology of Cr/CrN multilayer coating before and after erosion. It can be seen that after erosion, the Cr layer grains uniformly elongate along the interface and rotate by 90° compared with that before erosion. This change may be due to dislocation sliding between grains. The microstructure change of Cr layer is caused by plastic strain produced by erosion particles, while the microstructure (grain size and shape) of CrN layer is basically unchanged. Compared with the columnar structure, the microstructure composed of Cr/CrN multilayer grains grown at different interfaces can effectively prevent the propagation and propagation of cracks along the grain boundaries.

Therefore, the essence of multilayer structure to improve coating toughness is deeply explored [48]. Its toughening mechanism usually comes from the following aspects: (1) the interlayer interface deflects cracks; (2) The soft phase layer has better plastic deformation ability and can relieve the interface stress at the same time; (3) The crack tip is wrapped by a soft phase layer, which can passivate the crack tip, inhibit crack growth, and improve the toughness of the coating material (**Figure 7**). In simple terms, in multilayer coating structures, the metal phase is used to absorb

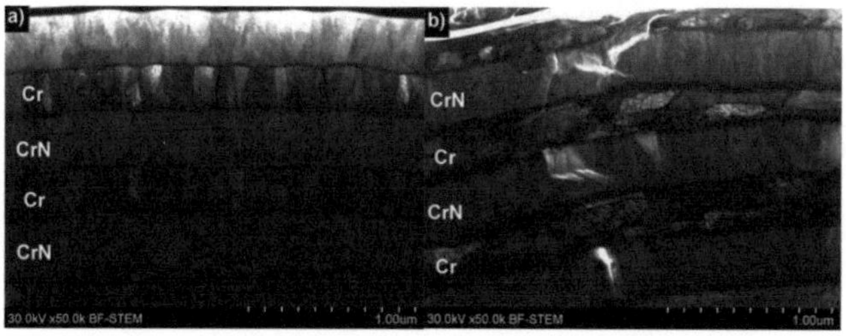


**Figure 6.**
*Microstructure of Cr/CrN multilayer coating (a) before erosion test, and (b) after erosion test [47].*

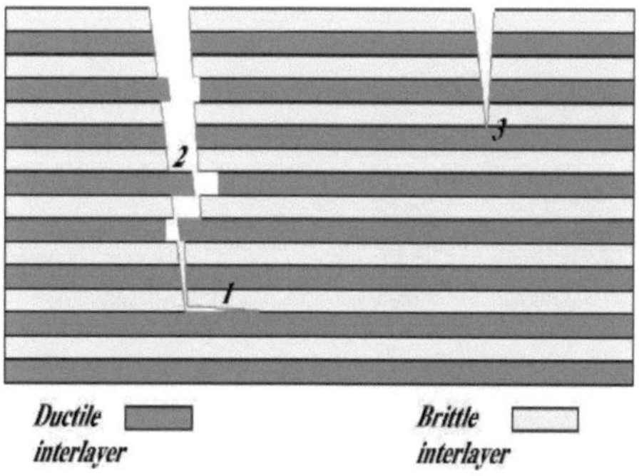


**Figure 7.**
*Schematic diagram of the toughening mechanism of multilayer film structure [48].*

excessive plastic deformation, while the ceramic phase provides hardness and wear resistance. It can be seen that the multilayer structure improves the toughness of the coating by a variety of mechanisms. These toughening mechanisms have been verified in a number of studies.

In general, the ductile interlayer improves the erosion resistance of the hard coating by improving the film substrate bonding strength and toughness. In the structural design of erosion resistant coatings, the multilayer structure of ductile interlayer/hard surface layer can be used for reference as the research direction of strengthening and toughening coatings.

## 5. Research and application status of SPE coating

In the past 30 years, the United States, Russia and other countries have cooperated with GE, Canada MDS, Liburdi and Russia PRAD, etc., and carried out research work in terms of material system, preparation technology and process, engineering test assessment and evaluation. Now, this technology has been successfully applied to dozens of models of engines.

As early as 1988, Liburdi Company in Canada began to develop erosion resistant coatings, and the coatings were gradually applied to various types of engines until the early twenty-first century. From 2000 to 2003, the anti-erosion coating prepared by the company was applied to T55, T58, T64 and AE1107 engines. In 2004, the company processed more than 2000 sets of T56 compressor blades with erosion resistant coatings for Rolls Royce. Since 2007, the Saudi and Jordanian Air Forces selected the T56 engine (compressor blades are provided with erosion resistant coatings prepared by Liburdi). In 2008, the third generation anti erosion coating (mainly TiAlN, as shown in **Figure 8**) was introduced, which can make the service life of the engine more than three times that of the uncoated engine, and improve the engine performance by 3% [44].

On the basis of the extensive application of binary anti-erosion coating, MDS-PRAD and GE further improved the material and structure of the anti-erosion coating, and applied ER-7 and Black Gold ceramic coatings, which are mainly composed of TiN and TiAlN, to helicopter and transport engine blades. The main component of ER−7 coating is TiN. It adopts a multilayer structure with alternating soft and hard. The substrate is a hard and dense nickel base metal. A transition layer between the substrate and the coating is used to improve the film substrate bonding strength. At the same time, the coating has strong resistance to fatigue crack growth and multi angle sand erosion.

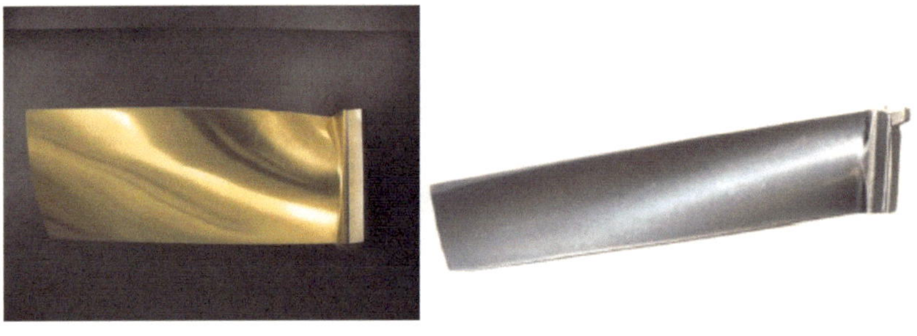

**Figure 8.**
*I & II generation coating (TiN series) III generation coating (TiAlN series) [44].*

Praxair Surface Technologies has developed a sub stoichiometric TiN/$TiN_{1-x}$ (called "24k Type II™") Compared with the traditional TiN coating, its erosion resistance has been greatly improved. This coating system has been tested to provide excellent gravel erosion protection in a variety of aircraft engines, including civil engines. Due to the increased demand for erosion resistant coatings in desert environments, the U.S. Navy has implemented a plan to extend the life of compressor blades of T64 helicopter engines by applying erosion resistant coatings [49, 50]. The company has established a production base, which can effectively prepare 24 k Type II™ Multilayer coating, and has become one of the major suppliers in aviation applications. Since then, the coating has been applied to hundreds of thousands of blades. The coating structure is shown in **Figure 9** $TiN_{1-x}$ layer thickness is less than 0.2 μm, TiN layer thickness is 1 μm. The total coating thickness is usually 15–25 μm. **Figure 10** shows the comparison diagram of erosion rate of multilayer coating and monolayer TiN coating, thermal spraying coating (chromium carbide and tungsten carbide) and Ti-6Al-4 V base material at 20° and 90° attack angles. The results show that the multilayer coating is obviously superior to other coatings [51].

Oerlikon Balzers has also provided BALINIT TURBINE PRO coating for erosion protection in recent years (the relevant performance is shown in **Figure 11**). This coating uses TiAlN series multilayer structure to achieve the best matching of high hardness and residual compressive stress, providing excellent erosion protection performance.

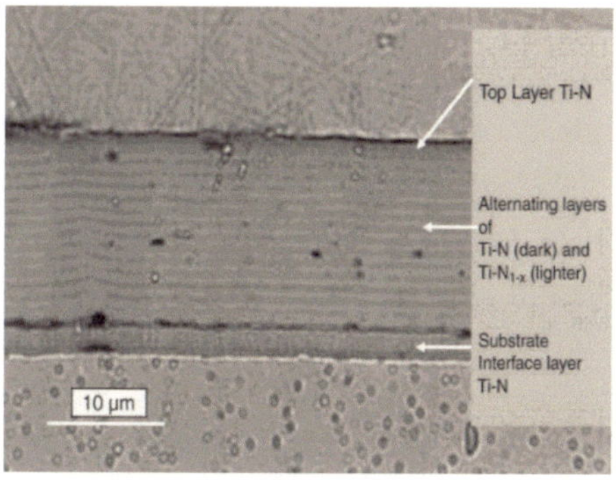


**Figure 9.**
*TiN/$TiN_{1-x}$ "24k Type II™" multilayer coating structure [51].*

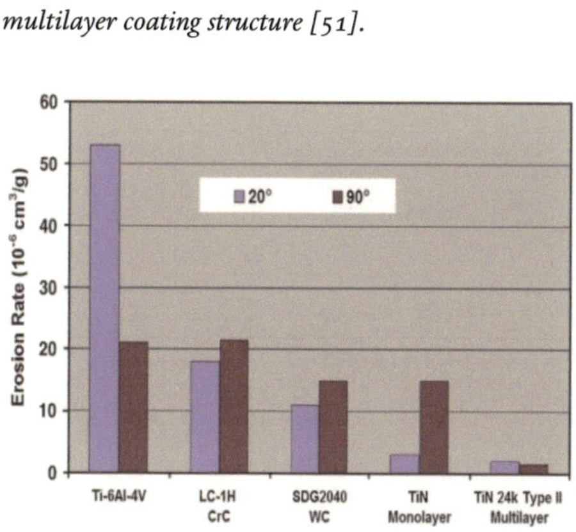


**Figure 10.**
*Erosion rates of Ti-6Al-4 V base material, thermal sprayed Cr-C and WC coating, TiN coating, and TiN/$TiN_{1-x}$ "24k Type II™" multilayer coating at attack angles of 20° and 90°, respectively [51].*

| | BALINIT TURBINE PRO |
|---|---|
| Coating material | MeAlN |
| Coating hardness $H_{IT}$ | 32 ± 2 GPa<br>4641 ± 300 ksi |
| Typical coating thickness (µm) | 5 – 25 |
| Friction against steel, dry running | ~0.5 |
| Coating temperature | < 500°C / < 932°F |
| Max. service temperature | 700°C / 1292°F |
| Colour | Violet-grey |

**Figure 11.**
*Basic properties of BALINIT TURBINE PRO coating.*

German MTU Company has designed and developed erosion resistant coatings (ERCoatnt Generation I and II) for aircraft engine compressor blades, as shown in **Figure 12**. **Figure 13a** shows the metallographic micro-section of the coating, clearly showing the total thickness of 25 µm and the ceramic and metal interlayer of about 3 µm per cycle. **Figure 13b** shows the high-resolution scanning electron micrograph of the coating, showing that the ceramic interlayer is a nano multilayer structure. The chemical composition of the continuous nano layers varies slightly, and the thickness of each nano layer is only 20 nm to 50 nm. Compared with traditional coatings, this nano design again significantly reduces the size of potential cracks or defects. In general, it is essentially chemical composition, and nano design and multilayer structure jointly achieve the ideal erosion resistance of the coating (**Figure 14**). **Figure 15** shows the effect of ERCoatnt coating on high cycle fatigue and low cycle fatigue

**Figure 12.**
*First-generation coating (TiN series) second-generation coating (TiAlN series) [40].*

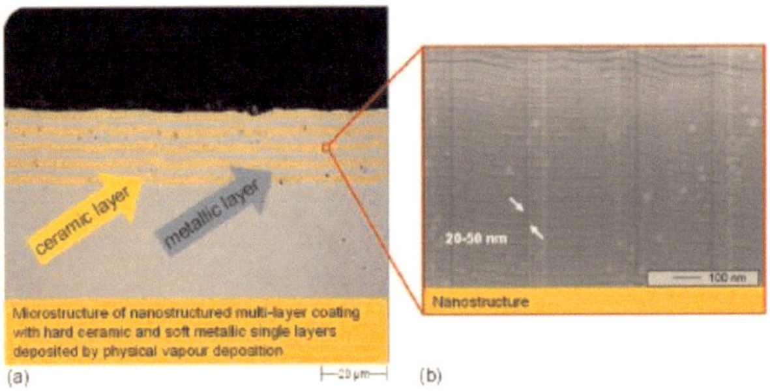


**Figure 13.**
*Multilayer structure of ERCoatnt coating [40]: (a) cross section (b) high resolution electron microscope cross section morphology.*

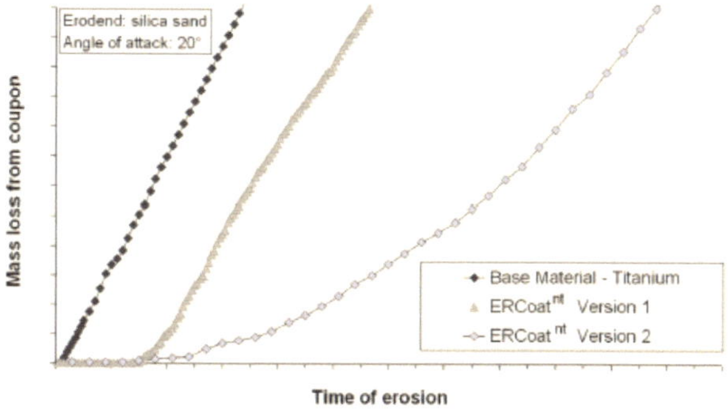


**Figure 14.**
*Erosion test results of coated and uncoated titanium samples [40].*

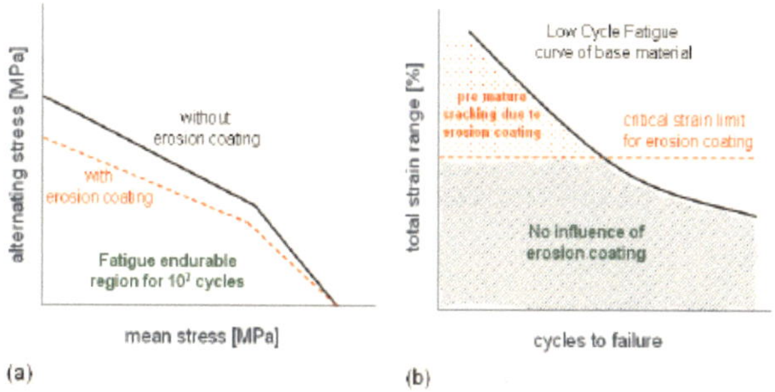


**Figure 15.**
*The effect of erosion-resistant coating on the fatigue strength of titanium alloy high-cycle (HCF) and low-cycle (LCF) (a) influence on HCF intensity (b) influence on LCF intensity [40].*

strength of titanium alloy. The test data showed that the high cycle fatigue strength decreased by as much as 15% depending on the coating system, the substrate material and the selected geometry. When the critical strain level is exceeded, ERCoatnt coating may accelerate crack initiation under low cycle fatigue load [40].

The team of Guangdong Academy of Sciences Institute of New Materials successfully developed 5–30 μm thick TiN based and CrN based alternating soft and hard multilayers by using vacuum cathodic arc ion plating technology [52–57]. The results show that: (1) The multilayers have good comprehensive properties: the thickness can reach 20 μm or more (see **Figure 16**), the adhesion is greater than 70 N, and the

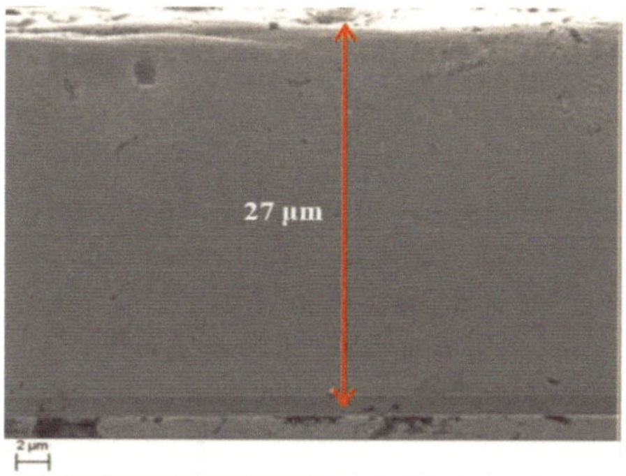


**Figure 16.**
*SEM cross-sectional morphology of Ti-TiN-Zr-ZrN multilayer film [52].*

hardness is greater than 30 GPa. It has a good anti-erosion protection effect on titanium alloy and steel base materials. (2) The coating prepared by arc has large metal particles which are not ionized, and these particles lead to a significant reduction in the overall performance of the coating, especially the erosion resistance. And (3) Due to the high hardness, the existing hard coatings form an "eggshell effect" with the titanium alloy substrate, which has a negative impact on the high stress low cycle fatigue performance of the substrate.

## 6. Problems and consideration of SPE coating

To sum up, in the research field of erosion resistant coatings, arc ion plating technology is widely used in the preparation of erosion resistant coatings for engine compressor blades. The coating system has gradually developed from binary TiN series coatings to multiple multilayer composite structure systems, and the comprehensive performance of the coatings has also been improved to varying degrees. However, up to now, the following problems still exist in the anti-erosion coating.

### 6.1 The internal influence law between composition, structure and performance of SPE coating

The composition selection and structure design of the coating are the key to determine its erosion resistance. At present, the research on erosion resistant coatings has been carried out to optimize the performance from the aspects of preparation parameters, coating microstructure, etc. The research results are one-sided and lack of systematicness and guidance. There is little research on the internal correlation between the composition structure and the residual internal stress, the interlayer interface and the erosion resistance of the coating. At the same time, there is a lack of in-depth study on the mechanism of coating erosion failure. In addition, according to the service condition requirements of the erosion resistant coating, it should also have high temperature resistance, fatigue resistance and other related properties. Therefore, based on the above performance requirements and actual service conditions, a complete anti-erosion coating design theory, such as material selection and structure design, should be established.

### 6.2 Matching of strength and toughness of SPE coating

A large number of research results have shown that the high hardness of the hard coating itself can well solve the micro cutting problem of the substrate caused by small attack angle erosion. However, due to the lack of toughness of the monolayer hard coating, the coating is prone to rapid failure in the form of brittle fracture when coping with the erosion of sand and gravel at high angles of attack. Therefore, the matching of strength and toughness of the coating is the key to breakthrough of the erosion resistant hard coating. How to obtain a hard and tough erosion resistant coating is also a research difficulty that has not been overcome so far.

### 6.3 Micro-particles in anti-erosion coatings prepared by arc ion plating

Arc ion plating technology has been widely used in the preparation of hard coatings. However, due to its physical characteristics of arc discharge, it is difficult to avoid the deposition of micro-particles of micrometer scale in the coating, which has a negative impact on the performance of the erosion resistant protective coating on the surface of precision parts. Therefore, how to eliminate or reduce the micro-particles in the anti-erosion coating prepared by arc ion plating is also one of the focuses of current research.

## Author details

Wang Di[1,2]* and Yang Zhen[2]

1 Shaanxi Key Laboratory of Surface Engineering and Remanufacturing, Xi'an University, Xi'an, China

2 Xi'an Flat Heat Treatment Co., LTD., Xi'an, China

*Address all correspondence to: wangd@xawl.edu.cn

## References

[1] Gachon Y, Ienny P, Forner A, et al. Erosion by solid particles of W/W-N multilayer coatings obtained by PVD process. Surface & Coatings Technology. 1999;**113**(1-2):140-148

[2] Wheeler DW, Wood RJK. Solid particle erosion of CVD diamond coatings. Wear. 1999;**233**(235): 306-318

[3] Yun-hong H, Yong-ming X, Shi-ting Y. Study on erosion behavior and mechanism of coating on steel structure in wind-sand environment. Journal of Tribology. 2010;**86**(01):26-31

[4] Junfeng B, In the moonlight, Bangs fly. Study on erosion wear mechanism of HVOF sprayed WC coating. Mining and Metallurgy. 2006;**15**(1):24-24

[5] Guanguo C, Xueti H. On the question of erosion wear. Journal of North China University of Technology (Natural Science Edition). 1997;**019**(004):27-32

[6] Dong H, Daoxin L, Yuntao X. Effect of shot peening on erosion resistance of solid particles in titanium alloy. Mechanical Science and Technology. 2008;**27**(004):466-470

[7] Lu B. Erosion-corrosion in oil and gas production. Chemical Engineering of Oil & Gas. 2013;**484**(4):473-493

[8] Finnie I. Some observations on the erosion of ductile metals. Wear. 1972; **19**(1):81-90

[9] Oh HL, Finnie I. The ring cracking of glass by spherical indenters. Journal of the Mechanics & Physics of Solids. 1967; **15**(6):401-406

[10] Bitter JGA. A study of erosion phenomena. Wear. 1963;**6**(1):5-21

[11] Wenwei X. Erosion Wear Simulation of Nitride Coatings. Xiangtan University; 2012

[12] Tilly GP. A two stage mechanism of ductile erosion. Wear. 1973;**23**(1):87-96

[13] Jiangai H, Yuwei W. Material Wear and Abrasion Resistant Material. Shenyang: Northeast University Press; 2001

[14] Levy AV, Chik P. The effects of erodent composition and shape on the erosion of steel. Wear. 1982;**89**(2): 151-162

[15] Sheldon GL, Finnie I. On the ductile behavior of nominally brittle materials during erosive cutting. Journal of Engineering for Industry. 1966;**88**(4): 387-389

[16] Sheldon GL, Kanhere A. An investigation of impingement erosion using single particles. Wear. 1972;**21**(1): 195-209

[17] Xiaoqing L. Study on Erosion and Wear Properties of High Strength Refractory Castable at Room Temperature. Xi'an: Xi'an University of Architecture and Technology; 2010

[18] Wood RJK, Wheeler DW, Lejeau DC, et al. Sand erosion performance of CVD boron carbide coated tungsten carbide. Wear. 1999; **233**(99):134-150

[19] Bose K, Wood RJK, Wheeler DW. High energy solid particle erosion mechanisms of superhard CVD coatings. Wear. 2005;**259**(1):135-144

[20] Deng J, Wu F, Lian Y, et al. Erosion wear of CrN, TiN, CrAlN, and TiAlN

PVD nitride coatings. International Journal of Refractory Metals & Hard Materials. 2012;**3**(5):10-16

[21] Dobrzanski LA, Polok M, Panjan P, et al. Improvement of wear resistance of hot work steels by PVD coatings deposition. Journal of Materials Processing Technology. 2004;**155**(156): 1995-2001

[22] Feuerstein A, Kleyman A. Ti-N multilayer systems for compressor airfoil sand erosion protection. Surface & Coatings Technology. 2009;**204**(6): 1092-1096

[23] Bromark M, Hedenqvist P, Hogmark S. The influence of substrate material on the erosion resistance of TiN coated tool steels. Wear. 1995;**187**(part-P1):189-194

[24] Swami V, Swaminathan P. Nanotechnology coatings for erosion protection of turbine components. Journal of Engineering for Gas Turbines and Power. 2010;**132**(8):82104-82104

[25] Wei R, Rincon C, Langa E, et al. Microstructure and tribological performance of nanocomposite Ti-Si-C-N coatings deposited using hexamethyldisilazane precursor. Journal of Vacuum Science & Technology A Vacuum Surfaces & Films. 2010;**28**(5): 1126-1132

[26] Immarigeon JP, Chow D, Parameswaran VR, et al. Erosion testing of coatings for aero engine compressor components. Advanced Performance Materials. 1997;**4**(4):371-388

[27] Oka YI, Yoshida T, Yamada Y, et al. Evaluation of erosion and fatigue resistance of ion plated chromium nitride applied to turbine blades. Wear. 2007;**263**(1):379-385

[28] Wang Q, Zhou F, Yan J. Evaluating mechanical properties and crack resistance of CrN, CrTiN, CrAlN and CrTiAlN coatings by nanoindentation and scratch tests. Surface & Coatings Technology. 2016;**28**(5):203-213

[29] Reedy MW, Eden TJ, Potter JK, et al. Erosion performance and characterization of nanolayer (Ti,Cr)N hard coatings for gas turbine engine compressor blade applications. Surface & Coatings Technology. 2011;**206**(2-3): 464-472

[30] Bin T, Xiaodong Z, Naisai H, et al. Study on the structure and tribological properties of CrN coating by IBED. Surface & Coatings Technology. 2000; **131**(1-3):391-394

[31] Chiba Y, Omura T, Ichimura H. Wear resistance of arc ion-plated chromium nitride coatings. Journal of Materials Research. 1993;**8**(5):1109-1115

[32] Banakh O, Schmid PE, Sanjinés R, et al. High-temperature oxidation resistance of $Cr_xAl_xN$ thin films deposited by reactive magnetron sputtering. Surface & Coatings Technology. 2003;**16**(3):57-61

[33] Ding XZ, Zeng XT. Structural, mechanical and tribological properties of CrAlN coatings deposited by reactive unbalanced magnetron sputtering. Surface & Coatings Technology. 2005; **200**(5-6):1372-1376

[34] Kim JW, Kim KH, Lee DB, et al. Study on high-temperature oxidation behaviors of Cr–Si–N films. Surface & Coatings Technology. 2006;**200**(24): 6702-6705

[35] Di W, Songsheng L, Lingyun L, et al. Effect of bias voltages on the microstructure and erosion resistance of CrAlN coatings deposited by arc ion

plating. Rare Metal Materials and Engineering. 2020;**49**(8):2583-2590

[36] Chunyan Y, Shebin W, Xiaoding Y, etc. Study on oxidation resistance of CrAlN films at high temperature. Rare Metals Materials And Engineering. 2009; **38**(006):1015-1018

[37] Endrino JL, Derflinger V. The influence of alloying elements on the phase stability and mechanical properties of AlCrN coatings. Surface & Coatings Technology. 2005;**200**(1-4): 988-992

[38] Endrino JL, Fox-Rabinovich GS, Gey C. Hard AlTiN, AlCrN PVD coatings for machining of austenitic stainless steel. Surface & Coatings Technology. 2006;**200**(24):6840-6845

[39] Fox-Rabinovich GS, Veldhuis SC, Scvortsov VN, et al. Elastic and plastic work of indentation as a characteristic of wear behavior for cutting tools with nitride PVD coatings. Thin Solid Films. 2004;**46**(9):505-512

[40] Xingrun R, Zhu H, Meixia L, et al. Comparison of microstructure and tribological Behaviors of CrAlN and CrN film deposited by DC magnetron sputtering. Rare Metal Materials and Engineering. 2018;**47**(4):1100-1106

[41] Fengfang W. Study on Erosion Wear Characteristics and Mechanism of PVD Nitride Coatings. Jinan: Shandong University; 2011

[42] Evans AG, Rosenblatt MEG. Impact damage in brittle materials in the elastic-plastic response Régime. Proceedings of the Royal Society of London. 1978; **361**(1706):343-365

[43] Hockey BJ, Wiederhorn SM, Johnson H. Erosion of Brittle Materials by Solid Particle Impact [M] Flaws and Testing. US: Springer; 1978

[44] Maurer C, Schulz U. Erosion resistant titanium based PVD coatings on CFRP. Wear. 2013;**302**(1-2):937-945

[45] Lishi W. Recent progress and prospect of ion plating hard film technology. Vacuum. 2000;**6**(4):1-11

[46] Xiao-Hua Z, Dao-Xin L. Effect of TiN/Ti multilayer on fretting fatigue resistance of Ti-811 alloy at elevated temperature. Transactions of Nonferrous Metals Society of China. 2009;**19**(003): 557-562

[47] Wieciński P, Smolik J, Garbacz H, et al. Erosion resistance of the nanostructured Cr/CrN multilayer coatings on Ti6Al4V alloy. Vacuum. 2014;**107**(6):277-283

[48] Berger M, Wiklund U, Eriksson M, et al. The multilayer effect in abrasion - Optimising the combination of hard and tough phases. Surface & Coatings Technology. 1999;**116**(119):1138-1144

[49] Engber M, Rüd K, Ardey S, et al. Advanced technologies for next generation regional jets survey of research activities at MTU aero engines. ISABE. 2007;**1282**:1-11

[50] Klein M, Simpson G. The development of innovative methods for erosion testing a Russian coating on GE T64 and T58 gas turbine engine compressor blades. Turbo Expo: Power for Land, Sea, and Air. 2004;**16**(5):42-51

[51] Dahotre NB, International A. Heat treating and surface engineering: Proceedings of the 22nd heat treating society conference and the 2nd international surface engineering congress. ASM International. 2003; **15**(17):453-462

[52] Lin SS, Zhou KS, Dai MJ, et al. Structural, mechanical, and sand erosion properties of TiN/Zr/ZrN multilayer coatings. Vacuum;**122**:179-186

[53] Di W, Lin S, Liu L, et al. Micro-nano multilayer structure design and solid particle erosion resistance performance of CrAlNx/CrAlN coating. Vacuum. 2020;**172**:109064

[54] Wang D, Lin S, Gong Y, et al. Solid particle erosion resistance of Cr-base gradient multilayer coatings. Surface & Coatings Technology. 2020;**402**:126352

[55] Wang D, Lin S, Shi Q, et al. Microstructure effects on fracture failure mechanism of CrAl/CrAlN coating. Ceramics International. 2021;**47**(3): 3657-3664

[56] Wang D, Lin S, Yang Z, et al. Failure mechanisms of CrN and CrAlN coatings for solid particle erosion resistance. Vacuum. 2022;**204**:111313

[57] Di WANG, Song-sheng LIN, Ling-yun LIU, et al. Effect of pulsed electromagnetic frequency on the microstructure, wear and solid erosion resistance of CrAlN coatings deposited by arc ion plating. Journal of Central South University. 2022;**29**:3065-3074

Chapter 2

# Periodic Instabilities in a Specific Low-Speed Pump Working as a Turbine

*Hernan Bolaños and Francisco Botero*

## Abstract

The hydrodynamic instabilities in a turbomachine can be divided into two main groups: periodic (or quasi-periodic) and nonperiodic. And the total instability, calculated from a statistical parameter with linear characteristics, such as variance, can be defined as the sum of periodic and nonperiodic instabilities. Based on the above, the main objective of the study was to estimate the periodic instabilities in a pump operating as a turbine. For this purpose, pressure fluctuation signals from sensors installed on the turbomachine volute and spaced 135° apart were used. The signals were analyzed in the time and frequency domain to identify, initially, the periodic instabilities and their relationship with the spectral components and, subsequently, to estimate the magnitude of these instabilities as the variance of the filtered series in the spectral band related to the periodic instability. In addition, the study aims to establish the contribution of periodic instabilities to total instability.

**Keywords:** pump as turbine, hydrodynamic instabilities, periodic instabilities, variance, phase analysis, Fourier analysis

## 1. Introduction

A reversible pump can either supply energy to the fluid or it can obtain energy from the fluid, depending on the direction of rotation of the impeller and the direction of flow. When the pump draws energy from the fluid, it is known as pump as turbine (PAT). The interest in this type of turbomachine is because power generation is less expensive in equipment than a conventional turbine for low power ranges [1–8]. Even though the use of PATs may be the best option for harnessing small hydro resources [9], especially in rural and remote areas with power supply problems [7, 8], the study of this type of turbomachines has not been extensive. This could be verified by a quick search on the Web of Science database, revised on November 1, 2022. There, 259,547 references are reported on pumps, 132,230 references on turbines, and only 281 references on PAT's. Of these 281 references, a general classification was made according to some keywords of interest in the field of turbomachines, which can be seen in **Table 1**.

| Query | Number of references |
|---|---|
| "Pump as turbine" | 281 |
| "Pump as turbine" AND performance | 203 |
| "Pump as turbine" AND performance AND efficiency | 151 |
| "Pump as turbine" AND performance AND design | 114 |
| "Pump as turbine" AND performance AND (cavitation OR "flow instabilities") | 24 |

**Table 1.**
*References on the web of science database, revised November 1, 2022.*

**Table 1** shows that the most studied issue related to PAT performance is efficiency, which is obvious given the importance of this aspect in power generation, and the least studied are cavitation and flow instabilities. Cavitation is an important source of instabilities, so it was included in the same query as instabilities. In this group, the following topics are addressed: geometry, efficiency, cavitation [4, 7–21], flow structures [8, 22–25], pressure fluctuations [16, 21, 26, 27], vortex rope [27], energy losses when switching from pump to turbine mode [28], and application of entropy production theory for energy losses [29]. None of them addresses the measurement of instabilities.

The study of instabilities is important because they cause several problems in the performance of turbomachines, such as efficiency losses, noise, and vibrations [12, 14, 16, 30–33] and can even threaten their structural integrity [16, 33]. Given this scenario, it is deemed important to identify the hydrodynamic phenomena that can affect the performance of a PAT and to estimate its level of instability. Brennen's classification [33] is used to establish a framework for flow instabilities in turbomachines. According to this author, hydrodynamic instabilities causing vibrations can be classified into three classes: global flow oscillations, local flow oscillations, and radial and rotodynamic forces. These three classes bring together at least 12 flow instabilities. The total instability at an operating point of a turbomachine may be the combination of several types of instabilities.

Instability analysis can be done using theoretical models; however, when the flows are very complex, it is necessary to use techniques that include experimental information [34]. One of those techniques is the frequency domain analysis of signals from sensors installed in the turbomachine. This is also known as Fourier analysis. An advantage of Fourier analysis is that the signals of the variables of interest obtained in the time domain can be converted into individual frequency components and vice versa. In addition, variables of interest (such as velocity, acceleration, and pressure, ), can be expressed as a sum of their mean and a complex component incorporating the amplitude and phase of the fluctuation [34]. In this paper, the words fluctuation, perturbation, and instability are synonymous.

Although the total instability at an operating point can be regarded as a sum of instabilities of different origins under the linearity assumption [34], not all these sources of instability have periodic characteristics. This means that, using Fourier analysis, only periodic or quasi-periodic instabilities can be clearly identified. Therefore, the main objective of this research is to estimate the instability due to periodic phenomena in a low-specific-speed pump working as a turbine. Concerning periodic instabilities, it is relevant to point out that they can be divided into two groups. The first group corresponds to those that depend on the rotation frequency of the turbomachine, and the second to those that do not. In this research, only those of the first group were considered.

## 2. Materials and methods

### 2.1 Test rig

The centrifugal pump under study, hereafter referred to as PAT, is an ITT Goulds of 1491.4 W (2 HP), with the specifications shown in **Table 2**.

The PAT was instrumented with pressure transducers in the high- and low-pressure orifices; two pressure fluctuation sensors on the volute; a torque sensor and an encoder on the shaft; a flow meter in the high-pressure pipe; and a frequency drive. The PAT was added to the test rig, which consisted of a closed pipe loop, a supply tank, and a recirculation pump, which was used to simulate the hydraulic head conditions for the PAT.

The pressure fluctuation sensors were placed on the volute in the same vertical plane, where sensor 1 (DYT1) and sensor 2 (DYT2) are 135° apart, as shown in **Figure 1**. The circular arrow inside the circle represents the direction of flow in the pump working as a turbine.

All sensor signals were acquired simultaneously by a National Instruments® CompaqRio® 9045 data acquisition unit. The signals from the pressure fluctuation, torque, and encoder sensors were acquired at a rate of 102.4 k samples/s. Signals from the pressure transducers and flow meter were acquired at 50 k samples/s.

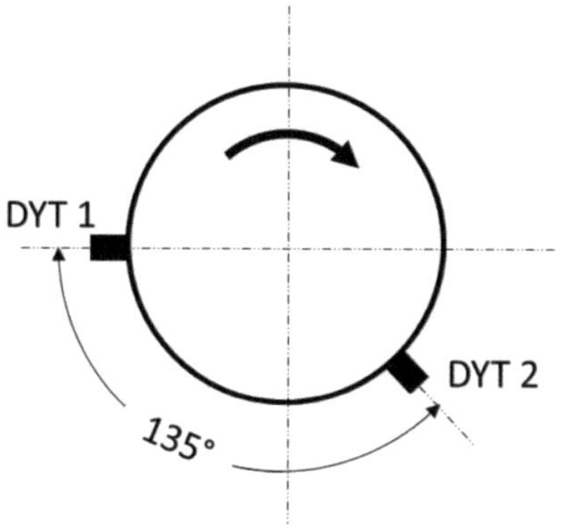


**Figure 1.**
*Diagram of the location of the pressure fluctuation sensors on the volute.*

| Parameter | Value |
|---|---|
| External diameter impeller | 0.14764 m |
| Reference diameter impeller | 0.08104 m |
| Number of blades (backswept) | 6 |
| Inlet diameter (low-pressure nozzle) | 0.0635 m |
| Outlet diameter (high-pressure nozzle) | 0.0508 m |

**Table 2.**
*Geometric characteristics of the PAT.*

## 2.2 $Q_{ED}$-$n_{ED}$ characteristic

The PAT characteristic curve was made based on dimensionless factors of velocity ($n_{ED}$) and discharge ($Q_{ED}$), according to IEC 60193 [35], as defined below:

$$n_{ED} = \frac{nD}{E^{0.5}} \tag{1}$$

$$Q_{ED} = \frac{Q}{D^2 E^{0.5}} \tag{2}$$

where $n$ is the rotation frequency ($s^{-1}$), $D$ is the reference diameter of the impeller (m), $E$ is the specific energy (N.m/kg), and $Q$ is the discharge ($m^3/s$). The characteristic curve was calculated with 60 operating points (OP), which had flow rates between $-0.0024$ and 0.0061 $m^3/s$, rotation frequencies between 5.04 and 33.85 $s^{-1,}$ and a hydraulic head around 8 m. The characteristic curve corresponds to quadrants 3 and 4 defined by IEC 60193 [35], with the operating modes given in **Table 3**.

## 2.3 Spectral analysis

Spectral analysis is performed to detect periodic or quasi-periodic components in a signal; therefore, it is important to differentiate these components from narrowband random contributions [36]. In practice, a simple way to identify periodic or quasi-periodic components is by means of a power density spectrum (PSD) or also known as an auto spectrum. In the PSD, the periodic or quasi-periodic component is identified as a sharp and clearly distinguishable peak. Other spectral representations derived from the PSD, such as the power spectrum (PS), linear spectrum (LS), and linear density spectrum (LSD) [37], can also be used to identify this type of component. In this work, PS was used to identify periodic components and PSD was used to measure the instability associated with hydrodynamic phenomena.

For the spectral analysis, we used a time series of 2,621,440 records (25.6 s) divided into 2.5 parts and overlapped by 50%. In this way, four segments were obtained, and the spectrum was estimated for each segment and then averaged. In addition, the Hanning window was used to fix possible discontinuities in the data series. The frequency resolution was 0.09765625 Hz.

### *2.3.1 Fourier analysis*

In turbomachines, Fourier analysis assumes that the fluctuations are periodic and linear [34]. With this assumption, instability can be considered as the sum of instabilities from different types or sources. In other words,

| Quadrant | Name | Operation mode | Remark |
|---|---|---|---|
| 3 | Turbine | Turbine | Discharge, velocity, and torque positives |
| | | Runaway | Discharge and velocity positives. Torque = 0 |
| | | Turbine brake | Discharge and velocity positives. Torque negative |
| 4 | Reverse pump | Reverse rotation | Velocity positive. Discharge and torque negatives |

**Table 3.**
*Operation modes in quadrants 3 and 4.*

$$\tilde{x} = \tilde{x}_1 + \tilde{x}_2 + \ldots + \tilde{x}_n = \sum_{i=1}^{n} \tilde{x}_i \quad (3)$$

where $\tilde{x}$ is the total instability of the variable $x$ and $\tilde{x}_1, \tilde{x}_2, \ldots, \tilde{x}_n$ are the instabilities by type or source. Hence, when passing from time domain to frequency domain, the spectral representation of the instability of the interest variable contains the spectral components of the instabilities by type or source. Thus,

$$\tilde{x} \xrightarrow{F} \tilde{X} \quad (4)$$

$$\tilde{X} = \tilde{X}_1 + \tilde{X}_2 + \ldots + \tilde{X}_n = \sum_{i=1}^{n} \tilde{X}_i \quad (5)$$

where Eq. (4) denotes the shift from time domain to frequency domain by means of the Fourier transform. $\tilde{X}$ is the frequency domain representation of the total instability of variable $x$ and $\tilde{X}_1, \tilde{X}_2, \ldots, \tilde{X}_n$ are the frequency domain contributions of instabilities by type or source. The Fourier transform estimation was performed by means of the Matlab® fft function.

### *2.3.2 Dimensionless representation of frequency and pressure fluctuations*

Frequency and pressure fluctuations were represented dimensionless. The frequency of the spectra was defined in terms of the frequency coefficient ($f_n$), according to IEC 60193 [35]:

$$f_n = \frac{f}{n} \quad (6)$$

where $f$ is the frequency of the spectral component ($s^{-1}$) and $n$ is the rotation frequency of the rotor or impeller ($s^{-1}$). The pressure fluctuation signals were represented by pressure fluctuation factor ($\tilde{P}_E$), which is defined by IEC 60193 [35] as:

$$\tilde{P}_E = \frac{\tilde{p}}{\rho E} \quad (7)$$

where $\tilde{p}$ is the pressure fluctuation ($N/m^2$), $\rho$ is the density ($kg/m^3$), and $E$ is the specific energy (N.m/kg).

### *2.3.3 Phase analysis*

The phase shift angle between corresponding spectral components was computed using the coefficients of the Fourier analysis. Phase analysis is useful to determine whether hydrodynamic phenomena are moving within the volute or whether they manifest as pressure pulses that simultaneously affect all parts of the system. In the first case, the phase difference is expected to be approximately equal to the angular separation between the sensors, and in the second case, approximately equal to zero. A negative phase lag means that the first signal is ahead of the second, and a positive phase lag is the opposite.

### 2.4 Measurement of flow instabilities due to low-frequency phenomena

The spectral analysis yields the spectral components or bands associated with hydrodynamic phenomena. To determine the flow instabilities, the signals are filtered into frequency bands containing the components of interest, and their instability in the time domain and then in the frequency domain is estimated on the filtered signals. The signals were filtered in the interest bands with digital FIR passband filters, designed with the Matlab® *designfilt* function. Filtering was performed on the signals from the pressure fluctuation sensors to determine the contribution of low-frequency phenomena to the total instability at each OP.

According to Hasmatuchi et al. [38], one measure of instability is the standard deviation. However, this statistic has the disadvantage that it does not have linear characteristics. That is, the standard deviation of the sum of the filtered signals is not equal to the sum of the standard deviations of each of the filtered signals. To overcome this limitation, the variance (Var) can be used as an indicator of the instability, since this statistic is linear in nature, and the following can be verified:

$$Var\left(\sum_{i=1}^{n}\tilde{x}_i\right) = Var(\tilde{x}_1) + Var(\tilde{x}_2) + \dots + Var(\tilde{x}_n) \tag{8}$$

where $\tilde{x}_1, \tilde{x}_2, \dots, \tilde{x}_n$ are the filtered signals in the frequency bands that were identified in the spectral analysis.

Considering that the area under the power density spectrum (PSD) is equal to the mean squared value of the signal [39], the variance of a zero-mean series can be used to measure the instability. This relationship between the area under the PSD and the variance of a zero-mean series allows for comparing the estimated instabilities in the frequency domain and in the time domain.

For estimation in the frequency domain, it would be enough to determine the area under the PSD in a frequency band associated with a component of interest, and this would be its approximate measure of instability. And as seen in Eq. (5), the linear nature of Fourier analysis allows the summing of the spectral power associated with the different phenomena to estimate the total instability.

To compare the estimated instabilities in the time and frequency domains, it is necessary to use the same units of measurement of the signals in the different domains. For this purpose, the technique proposed by Heinzel et al. [37] was used to convert the power spectral units into engineering units. In this research, the power spectrum is used since its representation in engineering units coincides with the units of variance.

## 3. Results

### 3.1 $Q_{ED}$-$n_{ED}$ characteristic curve

**Figure 2** shows the PAT operating modes in quadrants 3 and 4 of the $Q_{ED}$-$n_{ED}$ characteristic. The OPs corresponding to each operating mode are distinguished by colors. According to Greitzer [40], a hydraulic system can exhibit two forms of instability. The first, static instability, occurs when a small change in the flow rate at one OP can cause an increase in pressure forces and deviate the system to another OP farther away. The second, dynamic instability, occurs when a disturbance oscillates continuously increasing in amplitude. These forms of instability are associated with

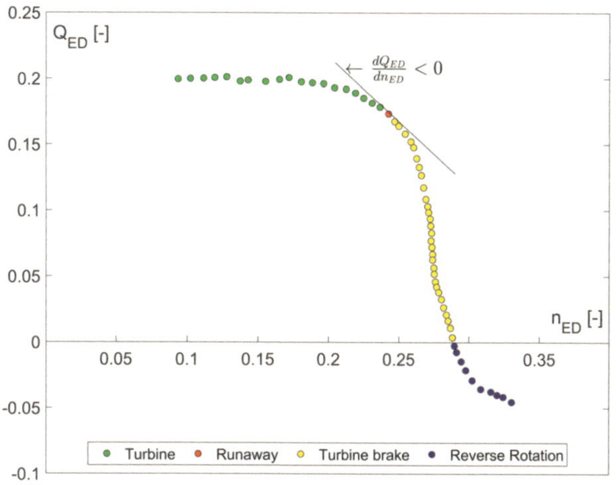


**Figure 2.**
*$Q_{ED}$-$n_{ED}$ characteristic.*

hydraulic transients and clues can be found by analyzing the $Q_{ED}$-$n_{ED}$ and $T_{ED}$-$n_{ED}$ characteristic curves [41–43]. The occurrence of static instability is a necessary but not sufficient condition for the occurrence of dynamic instability [43].

A practical method to identify static instability in the $Q_{ED}$-$n_{ED}$ curve is by the slope of a segment tangent to the runaway. If the slope is negative, the system is considered stable; if the segment is vertical, it is critical; and if the slope is positive, the system is unstable [41, 43]. The characteristic curve of an unstable system acquires the form "s," which means that for the same value of $n_{ED}$, several values of $Q_{ED}$ can occur. In some cases, the $Q_{ED}$ can reach values in all operating modes (turbine, turbine brake, and reverse pump). According to this criterion, the system under study is stable (see **Figure 2**) and, therefore, the system does not exhibit transient phenomena. This suggests that the instabilities occurring in the system are stationary.

## 3.2 Spectral analysis

### *3.2.1 Identification of periodic or quasi-periodic components*

This analysis was based on the waterfall spectra of the pressure fluctuation signals from DYT1 and DYT2 sensors (see **Figures 3** and **4**). In **Figures 3** and **4**, the spacing between the spectra (OP axis) is defined by the Euclidean distance between the OPs in the $Q_{ED}$-$n_{ED}$ plane. Comparing **Figures 3** and **4**, it is observed that there is a frequency correspondence in the components with higher spectral power, suggesting the existence of periodic or semi-periodic hydrodynamic phenomena detected by the two sensors. From this inspection, three spectral groups of interest were defined. Group 1 (G1) is composed of OPs from 32 to 48 in the band $0.76 \leq f_n \leq 0.86$, group 2 (G2) by OPs from 38 to 48 in the band $1.61 \leq f_n \leq 1.71$, and group 3 (G3) by all OPs in the band $5.95 \leq f_n \leq 6.05$. **Figures 3** and **4** also emphasize the groups of interest with shading.

G1 can be divided into two subgroups, with the following characteristics:

- Subgroup 1: It includes OPs 32–41 and describes a phenomenon that appears at OP 32, reaches its maximum instability at OP 38, and then decays up to OP 41. Its frequency coefficients vary between 0.7995 and 0.8378.

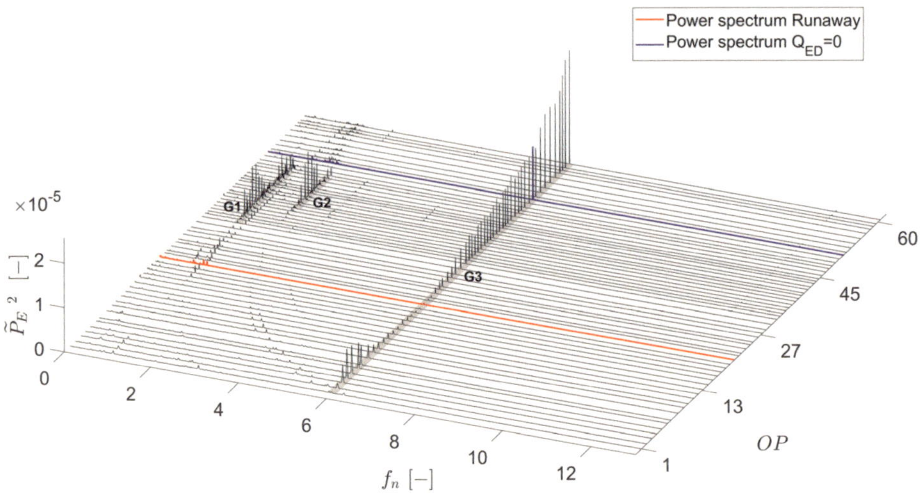


**Figure 3.**
*Waterfall power spectra for the DYT1.*

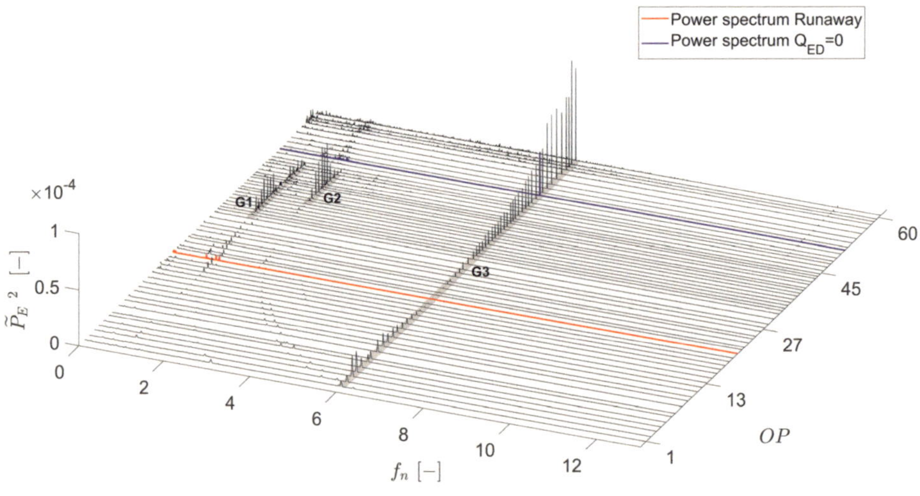


**Figure 4.**
*Waterfall power spectra for the DYT2.*

- Subgroup 2: It is constituted by OPs 42–48. It describes a phenomenon that initiates at OP 42, reaches its maximum instability at PO 44, and decays until it reaches OP 48. Its frequency coefficients vary between 0.7798 and 0.8381.

G2 contains OPs 38–48, where the way the spectral power is distributed suggests the same hydrodynamic phenomenon. The phenomenon begins at OP 38, increases in power up to OP 41, and from there decreases to OP 48. It occurs in the lower part of the turbine-brake mode, and its frequency coefficients range between 1.6438 and 1.6834.

G3 comprises all operating points and corresponds to the blade passage with $f_n = 6$. It is distributed throughout all operating modes in quadrants 3 and 4 and is most noticeable in turbine-brake and reverse pump modes. Near the runaway, the spectral power increases gradually until it reaches the last operating point in the reverse pump mode. In the high part of the turbine operating mode (OPs 3–9), some spectral powers are observed that outstand with respect to those of the other OPs in this part.

Another finding from comparing **Figures 2** and **3** is the spectral power of the components of interest in the DYT2 sensor signal spectra is higher than that of the DYT1 sensor. On average, the spectral powers in DYT2 are 2.6 times those of DYT1.

### *3.2.2 Characterization of periodic or quasi-periodic components*

The characterization was performed for the groups of interest under the hypothesis that the components are the representation of hydrodynamic phenomena detected by the two sensors. For this purpose, power spectra were used, and three criteria were applied for the selection of periodic components. First, components with sharp, clearly distinguishable, and isolated peaks [36], that is, without contiguous spectral components of similar spectral power. Second, corresponding components in the spectra of both sensors in terms of frequency factor, and third, coherence between them greater than or equal to 90%.

The characterization of G1 and G2 was performed by phase analysis. In G1, OPs 32, 33, 34, 36, 37, 38, 39, 44, 45, and 46 satisfy the periodicity criteria. These subsynchronous components have frequency coefficients between 0.8018 and 0.8378 and signal coherence between 90.3 and 99.7%. In OP 34, 37, 38, 39, and 45 the signal from DYT1 is ahead of DYT2, and they show phase shifts between −150.6 and −135.2°, approaching 135° when considering the separation of the sensors in the opposite direction of flow from DYT1. For OPs 32, 33, 36, 36, 36, 44, and 46, the phase shifts are between 206.3 and 236.9°, which are approximately 225°, the physical separation of the sensors in the opposite direction of flow but starting at DYT2. These components were subjected to an additional test, the wave number determination. This test serves to identify phenomena that sometimes decompose in rotating cells, such as rotating stall [44, 45]. For these phenomena, the wave number is equal to the number of rotating cells. The wave number was estimated as follows:

$$K = \frac{f_{n_{pg}}}{f_n} \tag{9}$$

where $K$ is the wave number, $f_{n_{pg}}$ is the passage frequency, and $f_n$ is the frequency identified in the PS. The passage frequency is the frequency perceived by the sensors and is estimated from the correlation between signals in a band containing the component of interest. Details of the method can be found in Refs. [44, 45]. For the periodic components of G1, it was found the passage frequency is approximately equal to the PS frequency, so the wave number is equal to one. The evidence collected suggests the existence of a one-cell subsynchronous hydrodynamic phenomenon (K = 1), moving around the volute in the opposite direction of flow. Full characterization of the phenomenon requires further studies that are beyond the scope of this investigation.

In G2, OP 39–47 fulfill the periodicity criteria. These components present frequency coefficients between 1.6438 and 1.6742, and coherence between 92.00 and 99.51%. OPs 39, 41, 43, 44, 46, and 47 present offsets between 108.1 and 161.8°, and OPs 40, 42, and 45 between −228.3 and −204.8°. Based on the same arguments used in the characterization of the periodic components of G1, it can be concluded the components of this group correspond to a one-cell phenomenon (K = 1) moving in the direction of the flow.

The G3 components correspond to instabilities due to rotor-stator interaction (RSI) blade excitation. In this case, the excitation is probably caused by the asymmetry of the volute [46] since the PAT has no guide vanes. The effect of the rotor blades on the volute produces a periodic disturbance or force whose frequency, expressed in terms of frequency coefficient, is given by:

$$f_{n_{S,k}} = Z_b k, \quad (k = 1, 2, 3, \dots) \tag{10}$$

where $f_{n_{S,k}}$ is the frequency coefficient of the disturbance due to the blades, $Z_b$ is the number of blades, and $k$ is harmonic. In our case, $f_{n_{S,1}} = 6$, is the blade passage.

Notice in **Figures 3** and **4** that OPs 20 (runaway) and 21 (next to runaway) do not show significant perturbations compared to the interest group components. On an appropriate scale, it can be seen that OPs 20 and 21 have a component that matches the periodicity criteria, with frequency coefficients of 0.776 and 0.758, respectively, coherence of 99.64% and 98.21%, and phase shifts of 0.1 and −3.3°, suggesting an in-phase phenomenon. This phenomenon is probably a surge, which is characterized by pressure and flow fluctuations that affect all parts of the hydraulic system simultaneously [33]. **Figures 5** and **6** show the PS of these points, where the highlighted spectral components are clearly distinguishable.

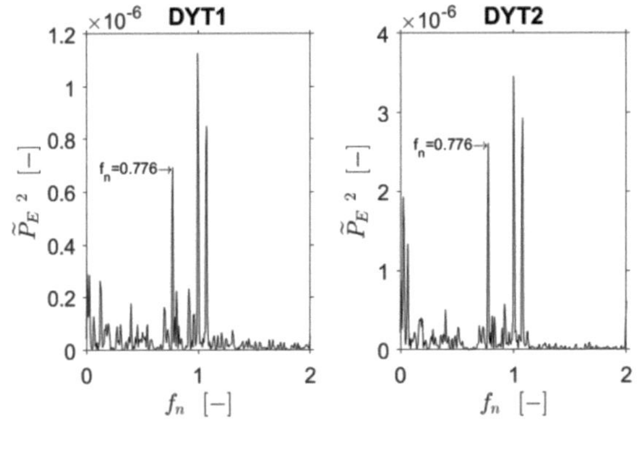


**Figure 5.**
*PS of OP 20 (runaway).*

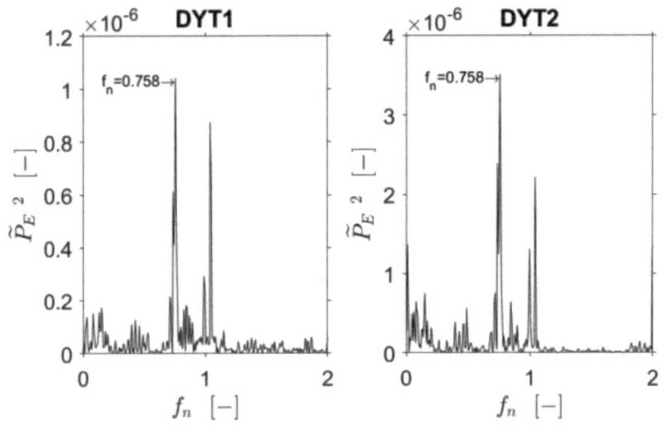


**Figure 6.**
*PS of OP 21 (next to runaway).*

### 3.3 Measurement of flow instabilities due to low-frequency phenomena

The instability measure of the prominent spectral components (G1, G2, and G3) at each OP was estimated based on the filtered signals at 2 Hz bandwidth. The bandwidth was equally distributed with respect to the spectral component of interest. In other words, the center of the filtering band corresponds to this component. Notice a distinction in the terms "interest band" and "filtering band." The former refers to the band defined by frequency coefficients ($f_n$) in which the spectral components of interest are found, and the latter refers to the frequency band in which the filtering was done around the prominent component identified in the interest band.

The filtered signals were used to estimate the instability due to periodic or quasi-periodic phenomena in the time domain and the frequency domain. In the time domain, we considered the filtered signals and computed the variance. In the frequency domain, we computed the PSD of the filtered signal and then estimated the area under the spectrum. **Figures 7** and **8** depict the three-dimensional representation of the instability estimates for the DYT1 and DYT2 sensor signals, respectively, based on the $Q_{ED}$-$n_{ED}$ characteristic curve and classified according to interest bands.

Comparing the estimates of instabilities in the time domain with the estimates in the frequency domain (**Figure 7a** vs. **b** and **Figure 8a** vs. **b**), it is observed they are not equal, but quite similar. Regarding the sum of variances, we obtain a mean error of 4.9% and a standard deviation of the errors of 13.5% for DYT1, and a mean error of 4.0% and a standard deviation of the errors of 11.7% for DYT2.

Looking at **Figures 7a** vs. **8a** (time domain instabilities estimation) and **Figures 7b** vs. **8b** (frequency domain instabilities estimation), it is suggested that the distribution of the pressure pulses is not homogeneous in the volute, since the magnitude of the instabilities in DYT1 is smaller than their corresponding ones in DYT2. If the most unstable operating zone is considered (OPs 32–48), the magnitudes of the instabilities in DYT2 are larger than their corresponding ones in DYT1 by 1.74–3.32 times.

In terms of the contribution of periodic or quasi-periodic phenomena to the total instability, DYT2 is analyzed, since the magnitude of instabilities perceived by this sensor are higher than those perceived by DYT1. In this case, the sum of the instabilities estimated with the filtered signals is compared with those of the original unfiltered signals, which were used to estimate the total instability. **Figure 9** shows a 3D representation of the sum of the periodic or quasi-periodic instabilities with respect to the total instability. Total instability estimations in the time and frequency domain yield a mean error of 0.3% with a standard deviation of the errors of 4.8%.

Given the contribution of periodic or semi-periodic instabilities to total instability (see **Figure 9**), three zones of operation can be defined. **Table 4** describes these zones.

## 4. Conclusions

The following conclusions can be drawn from the evidence collected:

At least four types of periodic or quasi-periodic phenomena exist in the PAT under study, which are described below:

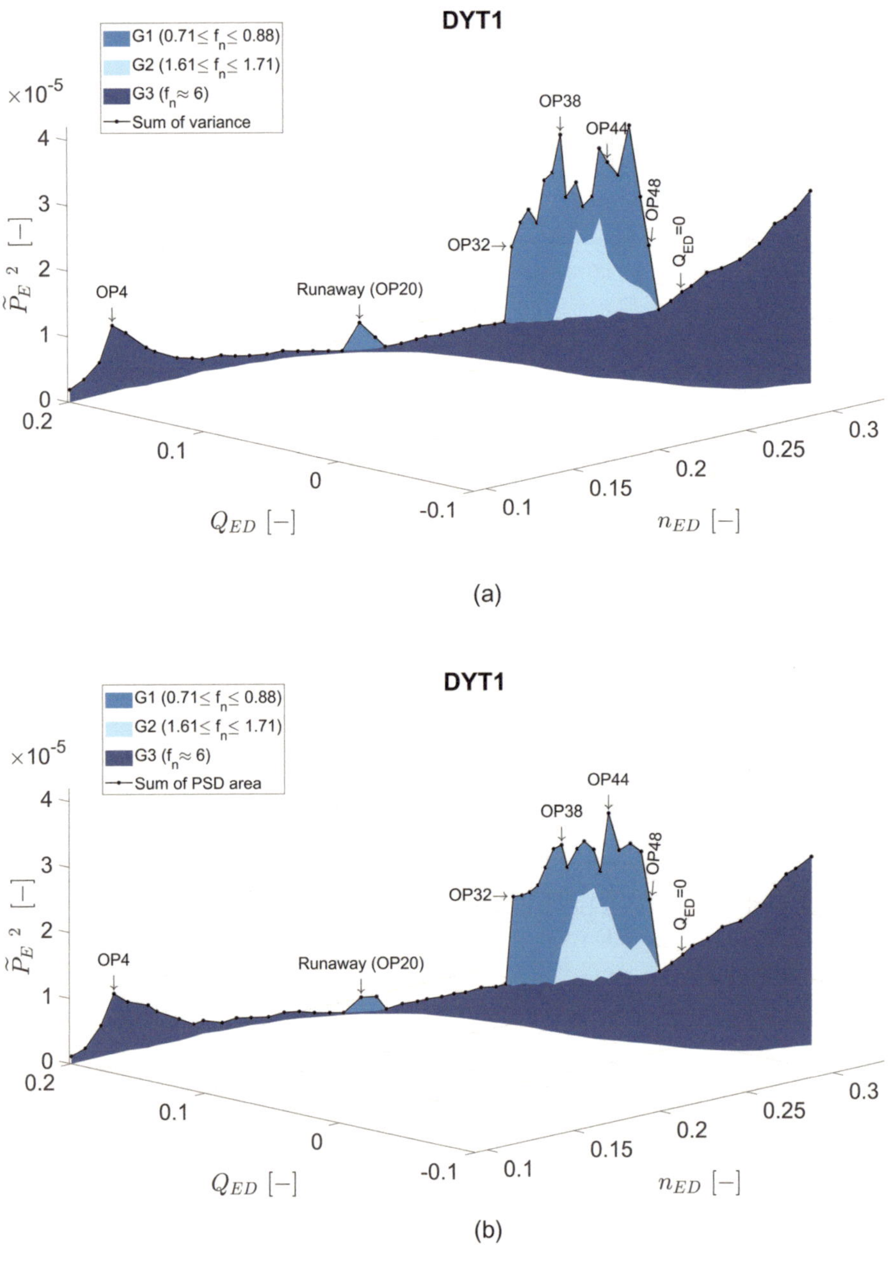


**Figure 7.**
*Estimation of periodic or quasi-periodic instabilities for DYT1. Above, estimation in the time domain. Below, estimation in the frequency domain.*

- In OPs 20 (runaway) and 21 (next to runaway), an in-phase phenomenon is detected, perhaps a surge with frequency coefficients of 0.776 and 0.758, respectively. Regarding the total instability, the disturbance magnitude of this phenomenon reaches 13.2 and 12.9% in DYT1 and 15.9 and 15% in DYT2, respectively.

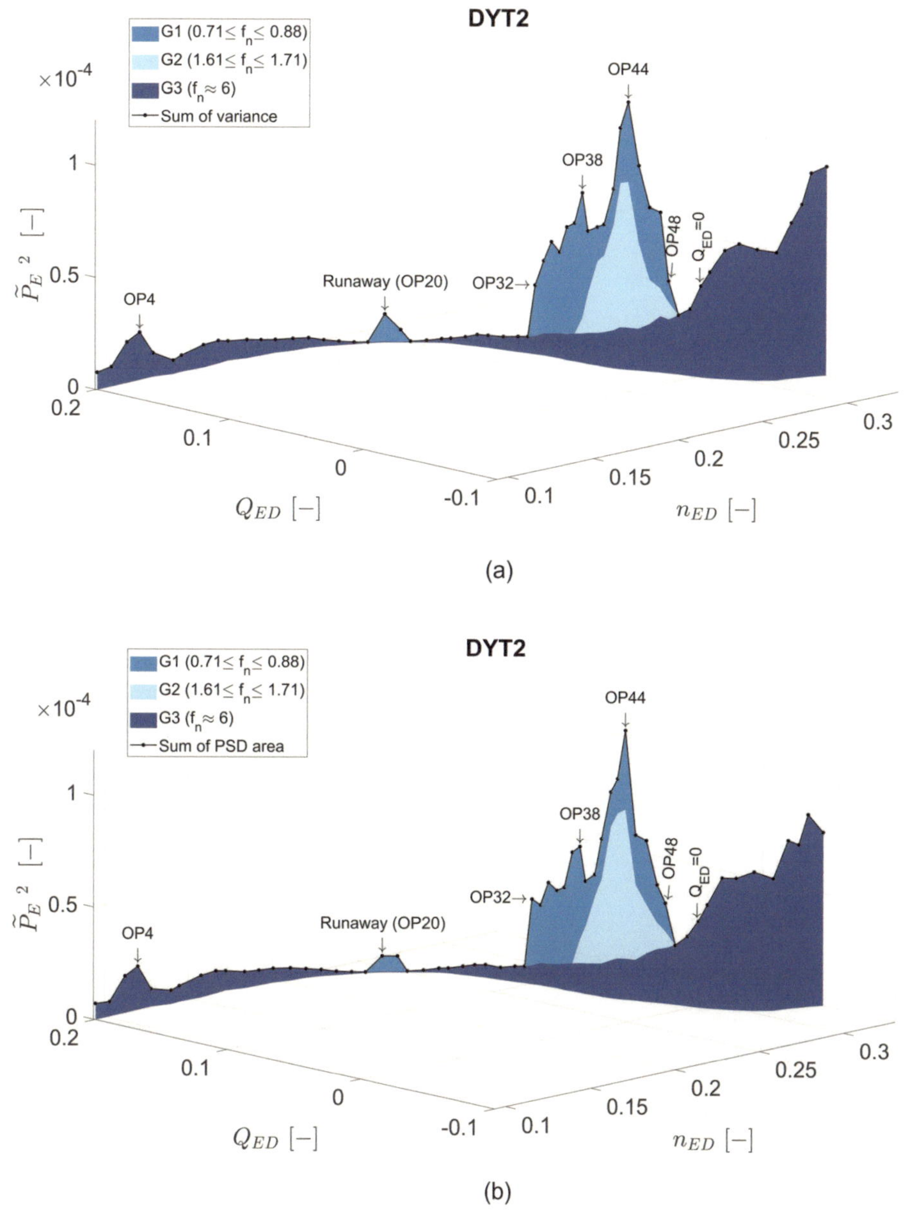


**Figure 8.**
*Estimation of periodic or quasi-periodic instabilities for DYT2. Above, estimation in the time domain. Below, estimation in the frequency domain.*

- At all operating points, blade excitation is present due to rotor-stator interaction, with $f_n = 6$. The magnitude of this disturbance gradually increases from the middle zone of the turbine-brake mode (OP 23) to the last OP in the reverse pump zone (OP 60). In the upper part of the turbine mode, there is a rapid increase and decay of this disturbance at OP 2–6. Between OP 18 and 22 the magnitude of the disturbance is practically negligible.

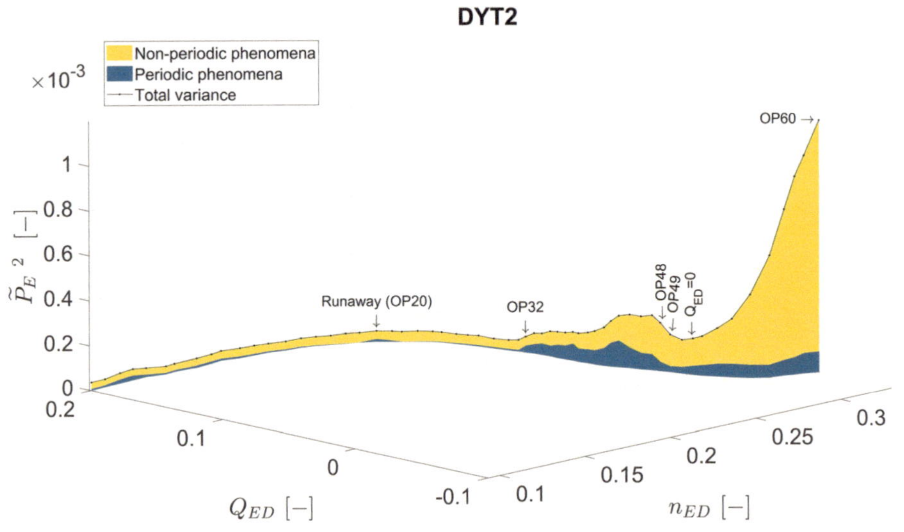


(a)

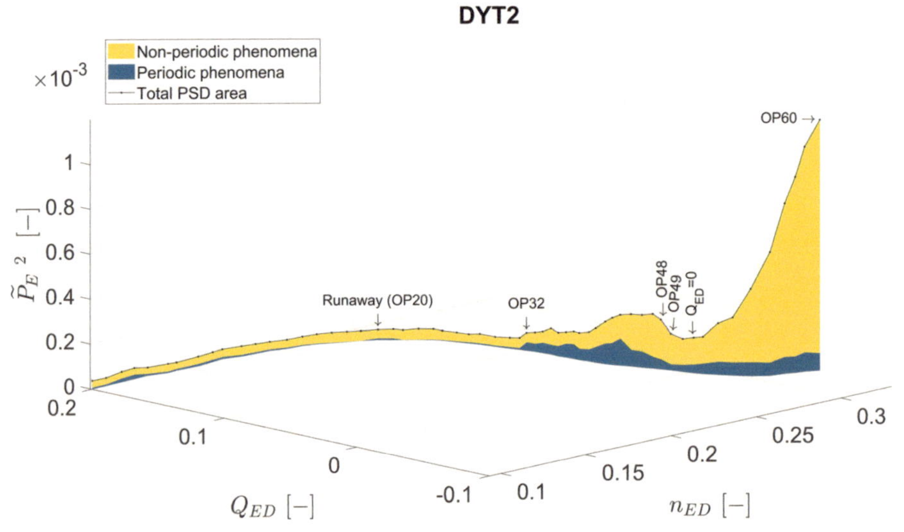


(b)

**Figure 9.**
*Contribution of periodic or quasi-periodic instabilities in the total instability for DYT2. Above, estimation in the time domain. Below, estimation in the frequency domain.*

- In G1, the evidence suggests the existence of a subsynchronous phenomenon that moves inside the volute in the opposite direction to the flow direction, with phase shift angles close to the physical separation of the sensors. In G2, a supersynchronous phenomenon is present and something like G1 occurs, only that the movement of the phenomenon is in the same direction of the flow.

In terms of the contribution of periodic instabilities to total instability, three zones can be identified. In the first one, constituted by OPs 1–31 (turbine, runaway, and

| Zone | OP | Remarks |
|---|---|---|
| 1 | 1–31 | Includes turbine, runaway, and upper turbine-brake modes. The total instability varies between $3.6\times10^{-5}$ and $5.3\times10^{-5}$ and is the least unstable operating zone of the PAT. The periodic phenomena' contribution to the instability varies between 1.2 and 42.2%, if the variance is considered, and between 1.1 and 39.2%, if the area under the PSD is considered. |
| 2 | 32–48 | It is in the lower part of the turbine-brake mode and at least three periodic phenomena are present, one of them fully characterized (blade excitation) and two partially characterized (see numeral 3.2.2). The total instability varies between $7.1\times10^{-5}$ and $2.5\times10^{-4}$ and is the area where the major contribution of the periodic phenomena to the total instability occurs, from 19.2 to 59.4%, if the variance is used, or from 20.1 to 54.7%, if the area under the PSD is used. |
| 3 | 49–60 | It is given from the lowest part of the turbine-brake mode and up to the last OP of the reverse pump mode. The total instability varies between $1.5\times10^{-4}$ and $1.1\times10^{-3}$. In this zone, blade excitation predominates and its contribution to the total instability is between 8.3 and 26.9%, if the variance is considered, or between 6.9 and 27.8%, if the area under the PSD is taken as a reference. |

**Table 4.**
*Operation zones in the PAT according to periodic or quasi-periodic instabilities contribution.*

upper part of turbine-brake), periodic instabilities represent approximately 1–42% of the total instability. In the second zone, between OP 32 and 48 (middle and lower turbine-brake mode), there are at least three hydrodynamic phenomena whose contribution to the total instability is between 20 and 60% approximately. The third zone covers OP 49–60 (last OPs in the turbine brake and all OPs of the reverse pump) where blade excitation predominates and presents contributions to the total instability ranging from 7 to 28%.

## Conflict of interest

The authors declare no conflict of interest.

## Author details

Hernan Bolaños* and Francisco Botero
EAFIT University, Medellín, Colombia

*Address all correspondence to: hbolanos@eafit.edu.co

## References


[1] Barbarelli S, Amelio M, Florio G, et al. Procedure selecting pumps running as turbines in micro hydro plants. Energy Procedia. 2017;**126**:549-556. DOI: 10.1016/j.egypro.2017.08.282

[2] Carravetta A, Derakhshan Houreh S, Ramos HM. Industrial aspects of PAT design improvement. In: Choi S, Duan H, Fu Y, et al., editors. Pumps as Turbines Fundamentals and Applications. Cham: Springer; 2018. pp. 59-73

[3] Gros L, Couzinet A, Pierrat D, et al. Complete pump characteristics and 4-quadrant representation investigated by experimental and numerical approaches. In: ASME-JSME-KSME 2011 Joint Fluids Engineering Conference, AJK 2011. 2011. pp. 359-368. DOI: 10.1115/AJK2011-06067

[4] Kan K, Binama M, Chen H, et al. Pump as turbine cavitation performance for both conventional and reverse operating modes: A review. Renewable and Sustainable Energy Reviews. 2022; **168**:112786. DOI: 10.1016/j.rser.2022.112786

[5] Wang W, Tai G, Shen J, et al. Experimental investigation on pressure fluctuation characteristics of a mixed-flow pump as turbine at turbine and runaway conditions. Journal of Energy Storage. 2022;**55**:105562. DOI: 10.1016/j.est.2022.105562

[6] Singh P, Nestmann F. Internal hydraulic analysis of impeller rounding in centrifugal pumps as turbines. Experimental Thermal and Fluid Science. 2011;**35**:121-134. DOI: 10.1016/j.expthermflusci.2010.08.013

[7] Xu J, Wang L, Asomani SN, et al. Improvement of internal flow performance of a centrifugal pump-as-turbine (pat) by impeller geometric optimization. Mathematics. 2020;**8**:1-23. DOI: 10.3390/math8101714

[8] Liu ZH, Yukawa T, Miyagawa K, et al. Characteristics and internal flow of a low specific speed pump used as a turbine. In: IOP Conference Series: Earth and Environmental Science. 2019;**240**: 042007. DOI: 10.1088/1755-1315/240/4/042007

[9] Amelio M, Barbarelli S, Schinello D. Review of methods used for selecting pumps as turbines (PATs) and predicting their characteristic curves. Energies. 2020;**13**:6341. DOI: 10.3390/en13236341

[10] Jain SV, Patel NK, Patel RN. Experimental investigations of cavitation characteristics of pump running in turbine mode. Journal of Energy Engineering. 2017;**143**:1-12. DOI: 10.1061/(asce)ey.1943-7897.0000387

[11] Ghorani MM, Sotoude Haghighi MH, Riasi A. Entropy generation minimization of a pump running in reverse mode based on surrogate models and NSGA-II. International Communications in Heat and Mass Transfer. 2020;**118**:104898. DOI: 10.1016/j.icheatmasstransfer.2020.104898

[12] Rakibuzzaman M, Jung KY, Suh SH. A study on the use of existing pump as turbine. In: E3S Web of Conferences. 2019;**128**:06004. DOI: 10.1051/e3sconf/201912806004

[13] Delgado J, Ferreira JP, Covas DIC, et al. Variable speed operation of centrifugal pumps running as turbines.

Experimental investigation. Renew Energy. 2019;**142**:437-450. DOI: 10.1016/j.renene.2019.04.067

[14] Graciano-Uribe J, Sierra J, Torres-López E. Instabilities and influence of geometric parameters on the efficiency of a pump operated as a turbine for micro hydro power generation: A review. Journal of Sustainable Development of Energy, Water and Environment Systems. 2021;**9**:1-23. DOI: 10.13044/j.sdewes.d8.0321

[15] Asomani SN, Yuan J, Wang L, et al. Geometrical effects on performance and inner flow characteristics of a pump-as-turbine: A review. Advances in Mechanical Engineering. 2020;**12**:1-19. DOI: 10.1177/1687814020912149

[16] Hao Y, Tan L. Symmetrical and unsymmetrical tip clearances on cavitation performance and radial force of a mixed flow pump as turbine at pump mode. Renewable Energy. 2018; **127**:368-376. DOI: 10.1016/j.renene.2018.04.072

[17] Zariatin DL, Rahmalina D, Prasetyo E, et al. The effect of surface roughness of the impeller to the performance of pump as turbine pico power plant. Journal of Mechanical Engineering Science. 2019;**13**:4693-4703. DOI: 10.15282/jmes.13.1.2019.24.0394

[18] Shi HX, Chai LP, Su XZ, et al. Performance optimization of energy recovery device based on pat with guide vane. International Journal of Simulation Modelling. 2018;**17**:472-484. DOI: 10.2507/IJSIMM17(3)443

[19] Derakhshan S, Kasaeian N. Optimization, numerical, and experimental study of a propeller pump as turbine. Journal of Energy Resources Technology. 2014;**136**:1-7. DOI: 10.1115/1.4026312

[20] Sengpanich K, Bohez ELJ, Thongkruer P, et al. New mode to operate centrifugal pump as impulse turbine. Renewable Energy. 2019;**140**: 983-993. DOI: 10.1016/j.renene.2019.03.116

[21] Han Y, Tan L. Influence of rotating speed on tip leakage vortex in a mixed flow pump as turbine at pump mode. Renewable Energy. 2020;**162**:144-150. DOI: 10.1016/j.renene.2020.08.033

[22] Li W. Vortex pump as turbine - a type turbine for energy generation or recovery based on computational fluid dynamics prediction. Journal of Fluids Engineering. 2019;**141**(10):101105. DOI: 10.1115/1.4042754

[23] Li W, Zhang Y. Numerical simulation of cavitating flow in a centrifugal pump as turbine. Proceedings of the Institution of Mechanical Engineers, Part E: Journal of Process Mechanical Engineering. 2018; **232**:135-154. DOI: 10.1177/0954408916686126

[24] Li WG, Zhang YL. Computational cavitating viscous liquid flows in a pump as turbine and Reynolds number effects. Proceedings of the Institution of Mechanical Engineers, Part E: Journal of Process Mechanical Engineering. 2019; **233**(3):536-550. DOI: 10.1177/0954408918770057

[25] Jianguo D, Adu D, Acheaw E, et al. Computational analysis on numerical simulation of internal flow physics for pump as turbine in renewable small hydro energy generation. Complexity. 2020:8869766. DOI: 10.1155/2020/8869766

[26] Zhang J, Appiah D, Zhang F, et al. Experimental and numerical investigations on pressure pulsation in a pump mode operation of a pump as

turbine. Energy Science & Engineering. 2019;7:1264-1279. DOI: 10.1002/ese3.344

[27] Yang F, Li Z, Yuan Y, et al. Study on vortex flow and pressure fluctuation in dustpan-shaped conduit of a low head axial-flow pump as turbine. Renewable Energy. 2022;**196**:856-869. DOI: 10.1016/j.renene.2022.07.024

[28] Kan K, Xu Z, Chen H, et al. Energy loss mechanisms of transition from pump mode to turbine mode of an axial-flow pump under bidirectional conditions. Energy. 2022;**257**:124630. DOI: 10.1016/j.energy.2022.124630

[29] Zhou L, Hang J, Bai L, et al. Application of entropy production theory for energy losses and other investigation in pumps and turbines: A review. Applied Energy. 2020;**318**: 119211. DOI: 10.1016/j.apenergy.2022. 119211

[30] Wang L, Asomani SN, Yuan J, et al. Geometrical optimization of pump-as-turbine (PAT) impellers for enhancing energy efficiency with 1-D theory. Energies. 2020;**13**:4120. DOI: 0.3390/ en13164120

[31] Berten S, Dupont P, Fabre L, et al. Experimental investigation of flow instabilities and rotating stall in a high-energy centrifugal pump stage. In: Proceedings of the ASME Fluids Engineering Division Summer Conference 2009, FEDSM 2009. 2009. pp. 505-513. DOI: 10.1115/ FEDSM2009-78562

[32] Brennen CE. A review of the dynamics of cavitating pumps. In: IOP Conf. Series: Earth and Environmental Science. 2012;**15**:012001. DOI: 10.1088/ 1755-1315/15/1/012001

[33] Brennen CE. Pump vibration. In: Hydrodynamics of Pumps. New York: Cambridge University Press; 2011. pp. 137-171

[34] Brennen CE. Unsteady flow in hydraulic systems. In: Hydrodynamics of Pumps. New York: Cambridge University Press; 2011. pp. 172-208

[35] International Electrotechnical Commission. Hydraulic Turbines, Storage Pumps and Pump-Turbines - Model Acceptance Tests (IEC 60193: 2019). London: BSI Standards Limited; 2019. p. 318

[36] Bendat JS, Piersol AG. Data acquisition and processing. In: Random Data Analysis and Measurement Procedures. Hoboken, New Jersey: John Wiley & Sons, Inc.; 2010. pp. 317-357

[37] Heinzel G, Rüdiger A, Schilling R. Spectrum and Spectral Density Estimation by the Discrete Fourier Transform (DFT). MPG Publication Repository. 2002. Available from: https://hdl.handle.net/11858/00-001M-0000-0013-557A-5

[38] Hasmatuchi V, Farhat M, Roth S, et al. Experimental evidence of rotating stall in a pump-turbine at off-design conditions in generating mode. Journal of Fluids Engineering. 2011;**133**:051104. DOI: 10.1115/1.4004088

[39] Bendat JS, Piersol AG. Basic descriptions and properties. In: Random Data Analysis and Measurement Procedures. Hoboken, New Jersey: John Wiley & Sons, Inc.; 2010. pp. 1-43

[40] Greitzer EM. The stability of pumping systems—The 1980 freeman scholar lecture. Journal of Fluids Engineering. 1981;**103**:193-242

[41] Olimstad G, Nielsen T, Brresen B. Stability limits of reversible-pump turbines in turbine mode of operation

and measurements of unstable characteristics. Journal of Fluids Engineering. 2012;**134**:111202. DOI: 10.1115/1.4007589

[42] Pejovic S, Karney B. Guidelines for transients are in need of revision. IOP Conference Series: Earth and Environmental Science. 2014;**22**: 042006. DOI: 10.1088/1755-1315/22/4/042006

[43] Zuo Z, Fan H, Liu S, et al. S-shaped characteristics on the performance curves of pump-turbines in turbine mode - a review. Renewable and Sustainable Energy Reviews. 2016;**60**: 836-851. DOI: 10.1016/j.rser.2015.12.312

[44] Best RC, Laflamme JGC, Moffatt WC. Flow measurements in rotating stall in a gas turbine engine compressor. Proceedings of ASME Turbo Expo. 1988;**1**:88-GT-219. DOI: 10.1115/88-GT-219

[45] Fortin J, Moffatt WC. Inlet flow distortion effects on rotating stall. Proceedings of the ASME Turbo Expo. 1990;**1**:90-GT-215. DOI: 10.1115/90-GT-215

[46] Dörfler P, Sick M, Coutu A. Periodic Effects of Runner-Casing Interaction. In: Flow-Induced Pulsation and Vibration in Hydroelectric Machinery. London: Springer-Verlang London; 2013. pp. 69-110

Chapter 3

# The Influences of Combined Distortions on Fan Performance

*Quanyong Xu, Anping Hou and Honghu Wu*


## Abstract

Combined distortions negatively affect the aerodynamic stability of fans, but previous studies have limitations. The 3D CFD method was used to calculate the aerodynamic stability of fans under the total pressure and swirl distortion, total temperature and swirl distortion, and total pressure and total temperature distortion in the work. Rotor 67 was taken as the research object. According to the calculation results, the influences of variable distortion parameters on fan performance under combined distortions were summarized to establish the functional relationship between stability margin losses and distortion parameters.


**Keywords:** combined distortion, stall, aerodynamic performance, fan, aero engine

## 1. Introduction

The inlet distortion is a common phenomenon in takeoff and flight. Some specific conditions, such as takeoff, aircraft inhaling missile exhausts, and aircraft use of the S-duct inlet, may lead to combined distortions at the aero-engine inlet, including the total pressure distortion, total temperature distortion, and swirl distortion. The inlet distortion negatively affects aero-engine performance, which directly threatens aero-engine stability and safety. Therefore, it is necessary to study the influences of combined distortions on aero-engine stability.

Researchers have studied the influences of combined distortions on the stability of the engine compression system for many years. They focus on the total pressure and total temperature distortion instead of other combined distortions. Braithwaite and Braithwaite *et al.* obtained the operating parameters of the engine under the 180° total pressure and total temperature distortion in different phases by testing engines TF30-P-3 and YTF34. The pressure distortion required to stall the compressor increases as the temperature distortion decreases for aligned high temperatures and low pressures. The required pressure distortion increases with the increased temperature distortion for the opposed high-temperature low-pressure case. The most persistent distortion occurs when the high temperature and low pressure are in the same 180° sector; the distortion attenuates the most when the low-pressure zone is opposite to the high-temperature zone [1, 2].

Mehalic studied the temperature and pressure distortion of the engine by testing engine PW1128. When the low-pressure zone coincides with the high-temperature zone, it will cause the high-pressure compressor to stall. When the low-pressure zone is opposite to the high-temperature zone, the influence on the stability of the high-pressure compressor will be weakened [3]. Davis used a stage-by-stage parallel compression system simulation and found the largest loss in stall-margin results under the steady-state pressure and temperature distortion when the high-temperature and low-pressure regions coincide. The stability limit has been lowered for the steady-state pressure distortion and temperature ramps. Stability is related to the position of the time for the dynamic combined distortion of the pressure oscillation and temperature gradient (the linear increase of temperature over time) when the temperature gradient is applied in the pressure oscillation cycle [4]. Modified parallel compressor theory can simulate more complex distortion patterns, including temperatures, pressures, and swirls. However, radial distortion effects cannot be included easily due to the changes in radial work distributions [5].

Frohnapfel *et al.* conducted experimental investigations and found that it is impossible to completely decouple total pressure and swirl distortions. A total pressure distortion typically creates a secondary flow field as air mixes from high to low-pressure regions within the distortion. Similarly, flow losses of axial flow at the inlet associated with turning under the swirl distortion for swirl distortion testing disturb the flow pressure field. The radial and tangential flow angles at the fan outlet are significantly influenced by the inlet-swirl distortion; whereas, the total and static pressure ratios of the fan are significantly influenced by the total pressure distortion at the inlet [6].

Huang found that the effect of combined distortions on engine stability is a complex nonlinear relationship rather than a simple addition of pressure and temperature distortion effects [7]. Ye *et al.* found that a compressor system creates maximum stability margin degradation when pressure distortion and temperature distortion phases coincide based on the numerical simulations of the 2D Euler equation, and the stability margin can be improved in some phases if a combined-distortion phase is changed suitably [8]. Xie *et al.* used the parallel compressor method and found that the influence of combined distortions on the fan and compressor pressure ratio is the same as that of the single-pressure distortion [9].

The parallel compressor model is often used in studying combined distortion. Pearson and McKenzie first proposed the parallel compressor theory in 1959 [10]. Reid [11], Mazzawy [12], and other researchers supplemented the model [13]. The theory shows that a compressor under the distortion can be treated as two or more sub compressors operating in parallel, and different levels of pressure- or temperature-distortions may be imposed upon their inlets, e.g., some take in distorted/undistorted air. Each sub compressor is assumed to operate independently of the other sub compressor except for the exit boundary where either a uniform static pressure or a uniform Mach number is imposed. As a result, each segment of the total system simulation will operate as a separate compression system. The entire compression system is considered to be unstable when one sub-compressor reaches the stability limit in this classical form. The mean operating point at instability is a weighted average of the low flow sector operating at the uniform flow stability boundary and the high flow sector operating at some other points far from the stability limit using the approach (see **Figure 1**) [14].

Research on the influences of combined distortions on the stability of the aero-engine compression system mainly adopts the parallel compressor method and the penetrating force method. However, the 3D computational fluid dynamics (CFD) numerical simulation is rarely used. Besides, research on combined distortions focuses

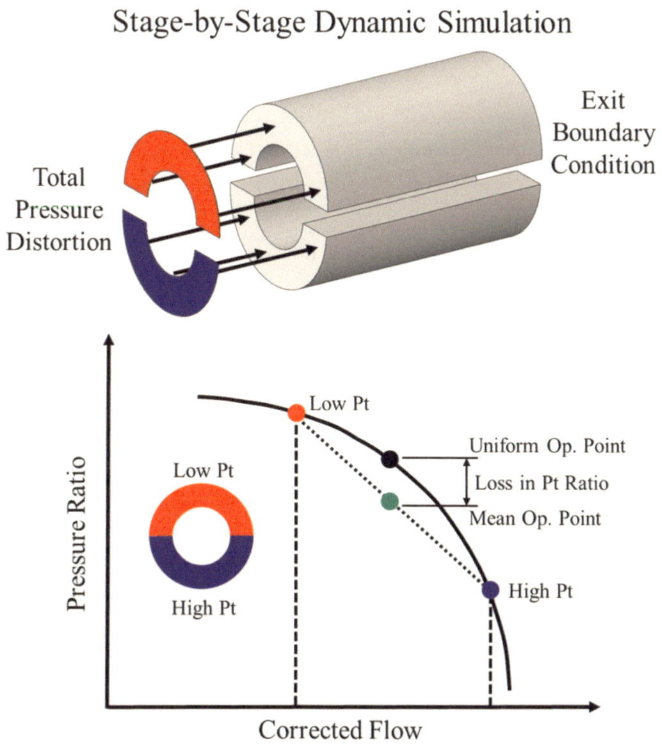


**Figure 1.**
*Parallel compressor theory [14].*

on the total pressure and total temperature distortion instead of combined distortions including the swirl distortion. The 3D CFD method was utilized to simulate the whole circumference of the rotor and study the influences of combined distortions on fan performance in the work. Valuable results were obtained.

## 2. Definition of inlet distortion parameters

The total pressure distortion, total temperature distortion, and swirl distortion parameters selected in the numerical simulation are as follows.

1. Steady-state circumferential-distortion parameter $\Delta\overline{\sigma}_0$ is selected as the total pressure distortion parameter. The index is the relative difference between the average total pressure recovery coefficient on the air interface plane (AIP) and the average total pressure recovery coefficient in the low-pressure zone. The low-pressure zone refers to the circumference range where the total pressure is less than the average total pressure on the AIP. The total pressure recovery coefficient is the ratio of the measured total pressure to the total pressure of undisturbed airflow before the inlet. The parameter is obtained by

$$\Delta\overline{\sigma}_0 = 1 - \frac{\sigma_0}{\sigma_{av}} \tag{1}$$

where $\sigma_0$ is the average total pressure recovery coefficient in the low-pressure region; $\sigma_{av}$ the face-average total pressure recovery coefficient at the AIP.

1. Relative increase of average surface temperature $\delta T_{2FAV}$ is selected as the total temperature distortion parameter. It numerically represents the temperature-distortion amplitude. The parameter is obtained by Eq. (2).

$$\delta T_{2FAV} = \frac{\Delta T_{2FAV}}{T_0} \tag{2}$$

   where $\Delta T_{2FAV} = T_{2FAV} - T_0$ is the increase of the face-average total temperature, K; $T_{2FAV}$ the face-average total temperature, K; $T_0$ the total temperature of the external test environment, K.

2. $\bar{\tau}_{87}$ is selected as the swirl distortion parameter. According to the preliminary calculation results, the inlet vortex is mainly the bulk swirl under the swirl distortion. $\bar{\tau}_{87}$ is the average $\tau_{87}$ value, which can represent overall vortex (bulk swirl) strength; $\tau_{87}$ is the ratio of the tangential component velocity at a radius of 0.87 $R_{max}$ to the axial flow rate. $\tau_{87}$ is the tangent of airflow tangential angle $\delta$ and is often directly expressed by the airflow tangential angle. **Figure 2** shows the installation of the $\tau_{87}$ detector and defines the tangential angle of airflow [15].

   The parameter is obtained by

$$\tau_{87} = \frac{v_t}{u} = \tan\delta \approx \delta \tag{3}$$

   where $u$ is the axial velocity, m/s; $v_t$ the tangential velocity, m/s;

3. Stability margin. We choose the stability margin evaluated at a constant corrected rotor speed (see Eq. (4)).

$$SM = \left[ \frac{\left( \pi_c / W_{a,c} \right)_{Su}}{\left( \pi_c / W_{a,c} \right)_{o}} - 1 \right] \times 100\% \tag{4}$$

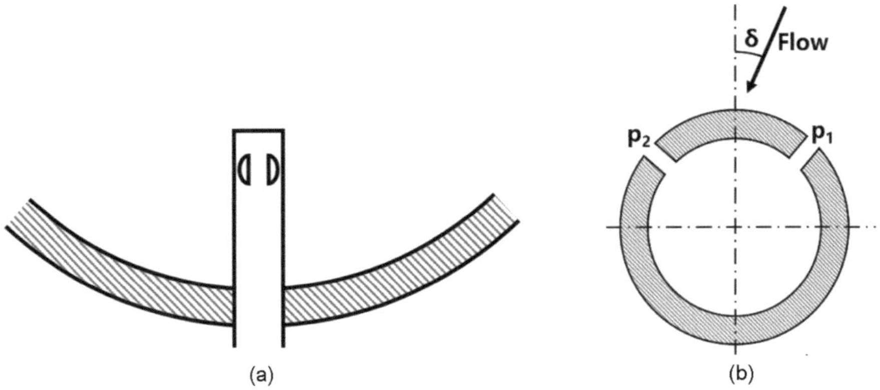


**Figure 2.**
*Definition of $\tau_{87}$ [15]. (a) Installation of the $\tau_{87}$ detector. (b) Definition of the airflow tangential angle.*

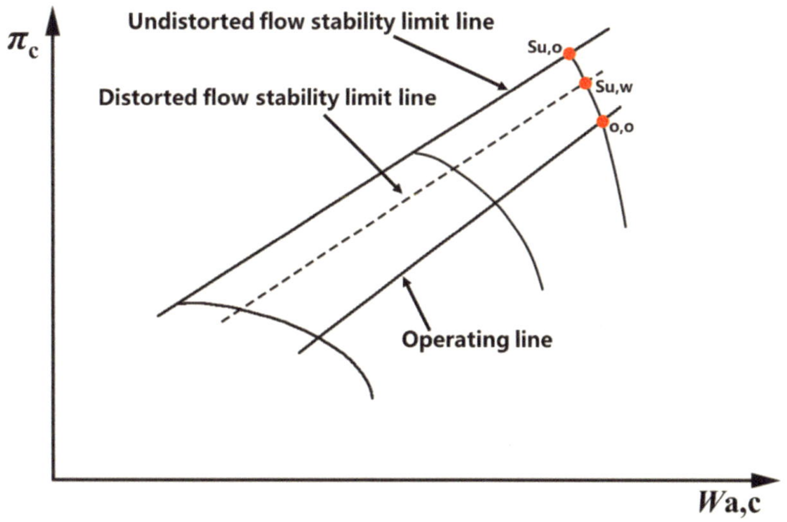


**Figure 3.**
*Stability margin definition.*

where *Su* is the stability-limit point, and *o* the operating point.

The stability margin loss is obtained by

$$\Delta SM_{Su,i} = SM_{Su,o} - SM_{Su,w} \tag{5}$$

where *Su, w* is distorted flow, and *Su, o* is undistorted flow (clean flow).

**Figure 3** shows the parameters in the definition of the stability margin loss.

## 3. Numerical simulation method

The numerical simulation of aircraft outflow is performed. The outlet section of the inlet duct in the aircraft outflow calculation results is taken as the inlet section in the internal-flow calculation of the compressor. Then, total pressure, total temperature, velocity component, and other gas parameters on the inlet section in the compressor are obtained by interpolation and combination according to the calculation conditions. Finally, the function of the stability margin loss and distortion parameters is established based on the calculation results.

The selected research object is NASA Rotor 67. Rotor 67 is a low-aspect-ratio transonic axial-flow fan rotor designed by NASA Lewis Research Center. The geometry of Rotor 67 and the experimental data obtained using both aerodynamic probes and laser anemometer measurements were reported by Strazisar *et al.* [16]. **Figure 4** shows the meridional view and 3D model of Rotor 67, and **Table 1** lists basic design specifications [17].

### 3.1 Numerical simulation method

The total pressure distortion, total temperature distortion, and swirl distortion should be combined to study the influences of combined distortions on the stability margin loss. The total pressure distortion, total temperature distortion, and swirl

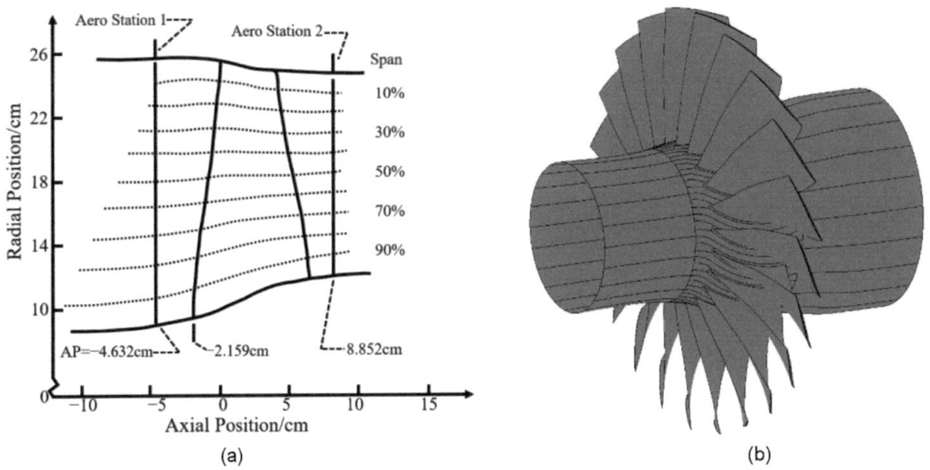


**Figure 4.**
*NASA rotor 67. (a) Geometry definition including stations [16]. (b) 3D model.*

| Specification | | Value |
|---|---|---|
| Number of rotor blades | | 22 |
| Rotational speed [rpm] | | 16,043 |
| Mass flow [kg/s] | | 33.25 |
| Total pressure ratio | | 1.63 |
| Rotor tip speed [m/s] | | 429 |
| Tip clearance at a design speed [cm] | | 0.061 |
| Inlet tip relative Mach Number | | 1.38 |
| Rotor aspect ratio | | 1.56 |
| Tip diameter [cm] | Inlet | 51.4 |
| | Exit | 48.5 |
| Hub-to-tip radius ratio | Inlet | 0.375 |
| | Exit | 0.478 |

**Table 1.**
*Basic specifications of NASA rotor 67 [17].*

distortion in the actual distortion of the outlet section of the inlet duct obtained in aircraft outflow calculation results all exist at the same time and change with the outflow conditions. Therefore, it is impossible to directly obtain the flow conditions with/without the swirl distortion or different swirl distortions under the total pressure and total temperature distortion. The inlet boundary of the rotor in the internal flow calculation is the outlet section of the inlet duct in the outflow calculation. The gas parameters at the outlet section of the inlet duct obtained from the outflow calculation should be interpolated and combined to obtain the internal flow calculation boundary conditions with adjustable distortion parameters (see **Figure 5**).

The total pressure distortion parameter is taken as an example. When the total pressure distortion parameter on the outlet section of the inlet duct in outflow

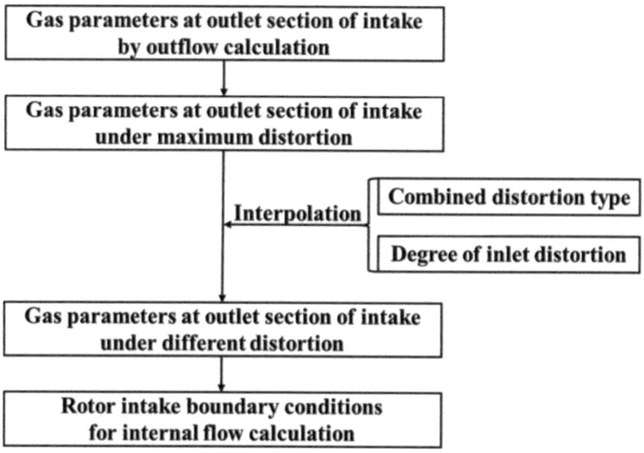


**Figure 5.**
*Interpolation and combination process.*

calculation results is the maximum, the total pressure at each node of the outlet section of the inlet duct under the corresponding working condition is taken as the reference value for pressure interpolations. Similarly, the reference values of the temperature interpolation and velocity-component interpolation can be obtained.

**Table 2** shows the corresponding working conditions of the maximum values of each distortion parameter. The pressure, temperature, velocity components, and other gas parameters at each node are interpolated in turn and then combined to obtain the required rotor inlet boundary conditions of the internal-flow calculation according to the relative distortion degree and the combined distortion type in the designed calculation condition. Besides, the interpolation of calculated grid nodes is required because different calculated grids correspond to the external conditions when the total pressure distortion parameter, total temperature distortion parameter, and swirl distortion parameter are maximum.

## 3.2 Verification of the numerical simulation method

Calculate the performance of Rotor 67 under undistorted intake air by the CFD method, and compare it with the experiment results of Rotor 67 [16] to verify the numerical simulation method. **Figure 6** shows the comparison between the characteristic maps obtained by the CFD calculation and experiment results. The pressure-ratio characteristic line of CFD calculation results is lower than that of experiment results under same airflow, and the characteristic line of the efficiency is relatively consistent. The comparison results show that the numerical simulation method used in the work can better fit the actual working conditions, with reliability.

| Distortion type | Maximum distortion parameter |
|---|---|
| Total pressure distortion | 5.7540% |
| Total temperature distortion | 18.7302% |
| Swirl distortion | 0.185257 |

**Table 2.**
*Corresponding working conditions of the maximum values of each distortion parameter.*

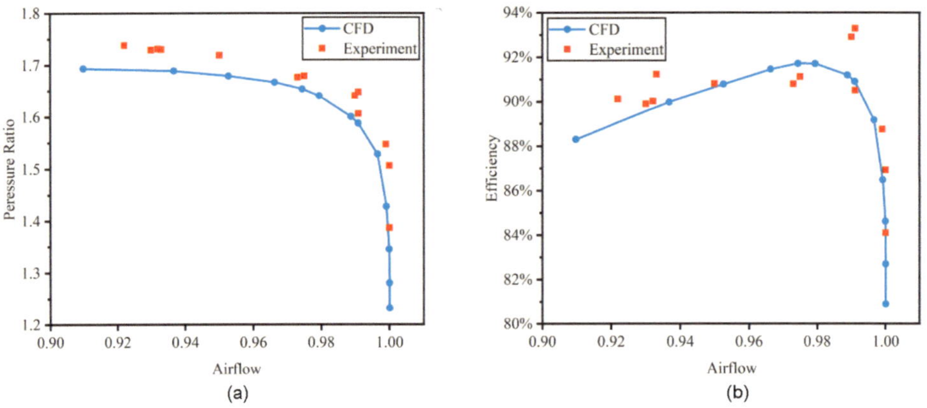


**Figure 6.**
*Comparison of characteristics between CFD results and experiment results. (a) Total pressure ratio. (b) Efficiency.*

# 4. Effect of pure distortions on aerodynamic stability of fans

## 4.1 Effect of the pure total pressure distortion on aerodynamic stability of fans

The distortion screens of variable solidity in the inlet duct block a 180° sector of the inlet annulus to generate the total pressure distortion of the compressor inlet. **Figure 7** shows typical circumferential profiles resulting from averaging the pressures on each rake. The total pressure profile reasonably approximates a 180° extent square wave at the compressor inlet, whereas the static pressure profile appears sinusoidal.

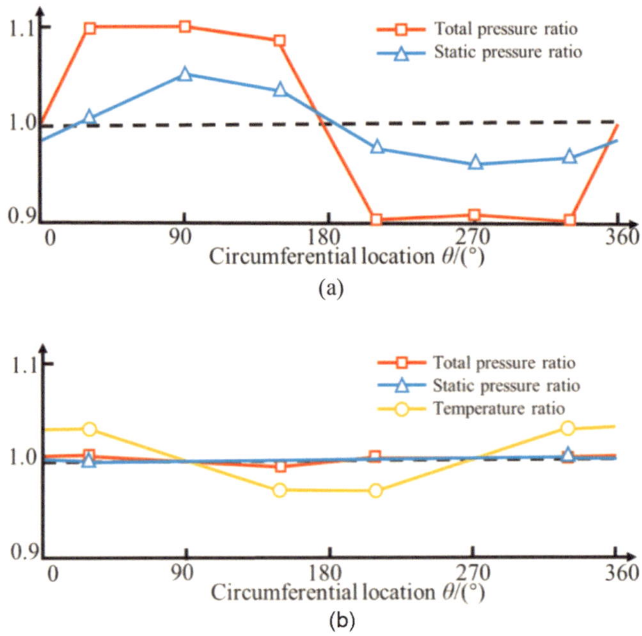


**Figure 7.**
*Compressor profiles with the 180° circumferential pressure distortion at the 94% designed speed [18].*
*(a) Compressor inlet. (b) Compressor outlet.*

The compressor-outlet profiles show very little circumferential variation in the total/static pressure; however, a total temperature distortion is evident. If the region of higher exit temperature is the result of the higher compression heat in the distorted sector, distorted flow rotates approximately 90° in the rotor-rotating direction passing from the low-pressure sector of the compressor inlet to the compressor outlet [18].

The effects of different distortion amplitudes (screen solidity) on the compressor-stall line are shown on the undistorted compressor map (see **Figure 8**). Each stall line is defined as the locus of stall points generated. $P_d$ is the average of the pressure measurements on the rakes behind the distortion screen; $P_{ud}$ is that in the undistorted region; $P_{av}$ is the average of $P_d$ and $P_{ud}$. The increased amplitude of the pressure distortion causes the stall line to move to lower pressure ratios. Also, the corrected rotor speed lines shift to the left or lower airflow for 93% of the corrected speed. The pressure distortion is not constant with airflow. The pressure distortion for screen "A" varies from 5% at 90% of the speed to 12% at the rated speed [18].

**Figure 9** shows the effects of different total pressure distortions on $\Delta(PRS)$. Parameters $\Delta(PRS)_W$ and $\Delta(PRS)_N$, describing losses in the pressure ratio of stall at constant corrected airflow or a constant corrected speed, are used to evaluate the effects of flow distortions. $\Delta P/P$, the ratio of $\Delta(PRS)_N$ to the amplitude of the distortion, is approximately 0.60 and essentially independent of speeds with inlet pressure distortion alone. The ratio of $\Delta(PRS)_W$ to $\Delta P/P$, however, varies with speeds and amplitudes from 0.50 to 1.0, and the ratio is constant with the amplitude only at 93% of the corrected speed [18].

### 4.2 Effect of the pure total temperature distortion on aerodynamic stability of fans

The gaseous hydrogen burner in the inlet duct heats a 180° sector of the inlet annulus to generate the total temperature distortion of the compressor inlet. **Figure 10** shows typical circumferential profiles resulting from averaging the pressure and

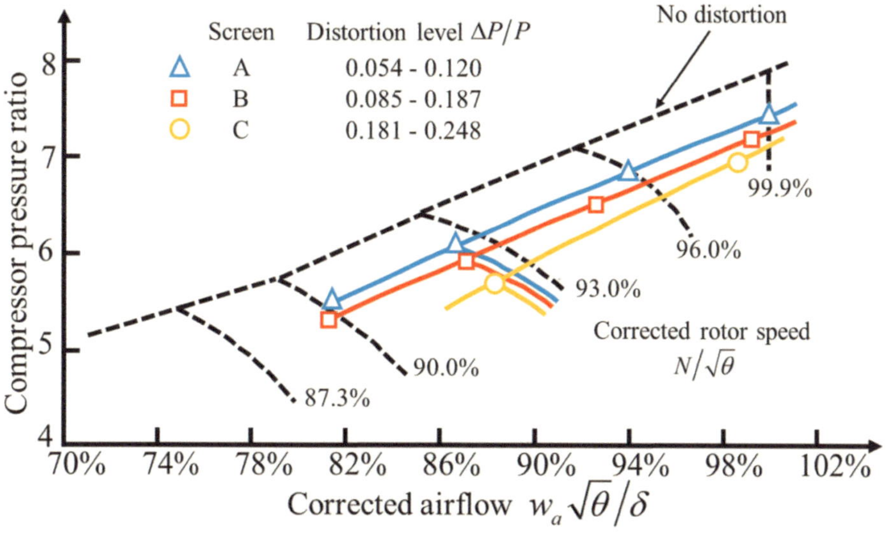


**Figure 8.**
*Compressor map showing the effects of the 180° circumferential pressure distortion [18].*

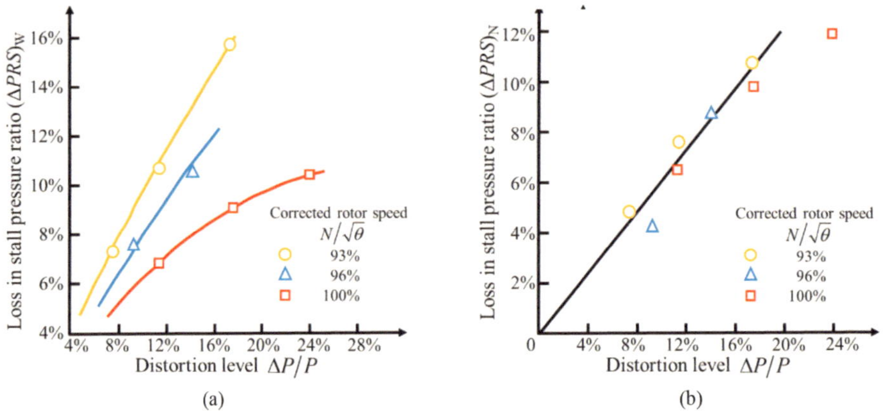


**Figure 9.**
*Losses in the pressure ratio of stall for the 180°-extent inlet-pressure distortion [18]. (a) At constant corrected airflow. (b) At a constant corrected speed.*

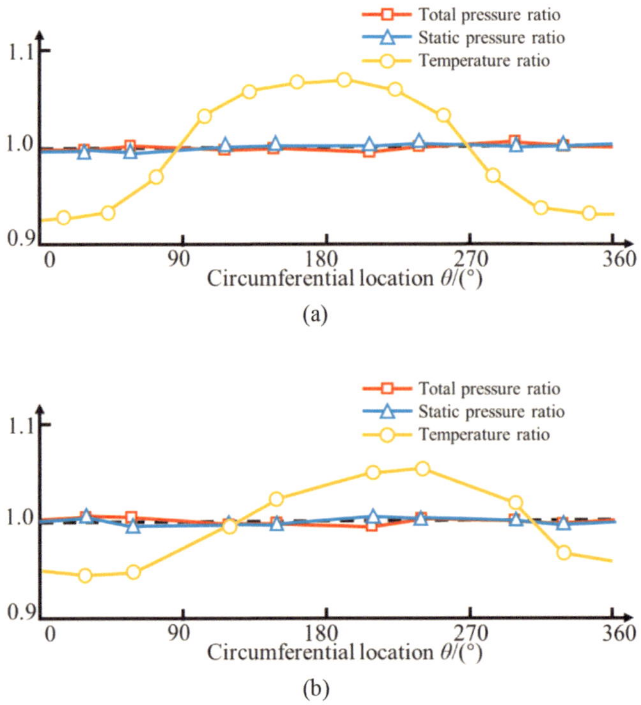


**Figure 10.**
*Compressor profiles with the 180° circumferential temperature distortion at 96% of the designed speed [18]. (a) Compressor inlet. (b) Compressor outlet.*

temperature on each rake. The temperature profile is more sinusoidal than the desired square wave pattern at the compressor inlet. The comparison of total temperature profiles of the compressor inlet and outlet indicates that the temperature distortion persists through the compressor with only a 29% reduction in the maximum amplitude, while the total pressure distortion is completely attenuated at the compressor outlet. Also, the peak temperature distortion rotates approximately 45° in the rotor-

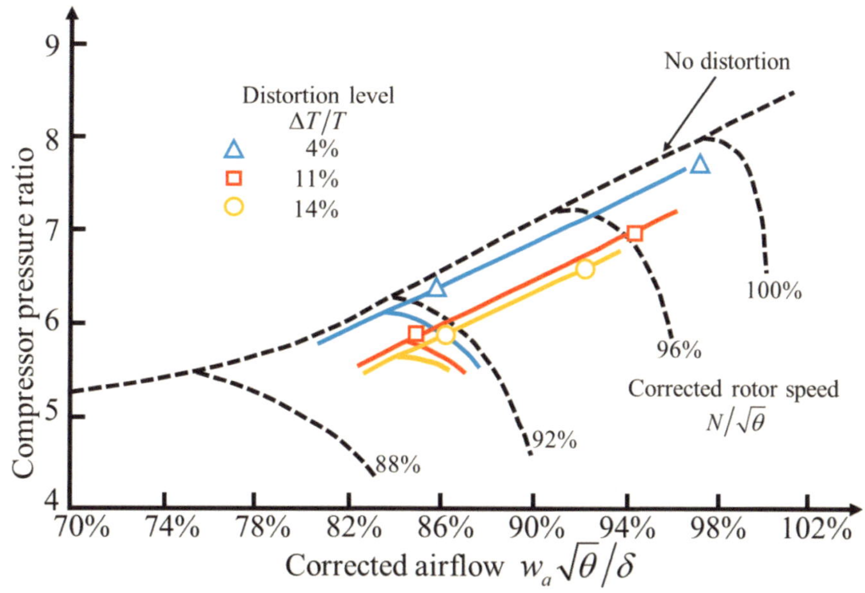


**Figure 11.**
*Effects of the 180° circumferential temperature distortion [18].*

rotating direction. There is little or no circumferential distortion of the total/static pressure associated with the temperature distortion at the compressor inlet or exit [18].

Effects on the stall line of different temperature distortions are presented on the undistorted compressor map (see **Figure 11**). $T_d$ is the average of the temperature measurements in the heated region; $T_{ud}$ is that in the undistorted region; $T_{av}$ is the average of $T_d$ and $T_{ud}$. The increased temperature distortion decreases the stall pressure ratio, while another effect is the decrease in the corrected rotor speed for a constant mechanical speed. The constant corrected speed lines also move to lower values of corrected airflow for 92% of the corrected speed [18].

**Figure 12** shows the loss in the pressure ratio of stall evaluated at constant corrected airflow $\Delta(PRS)_W$ under different total temperature distortions. It is nearly

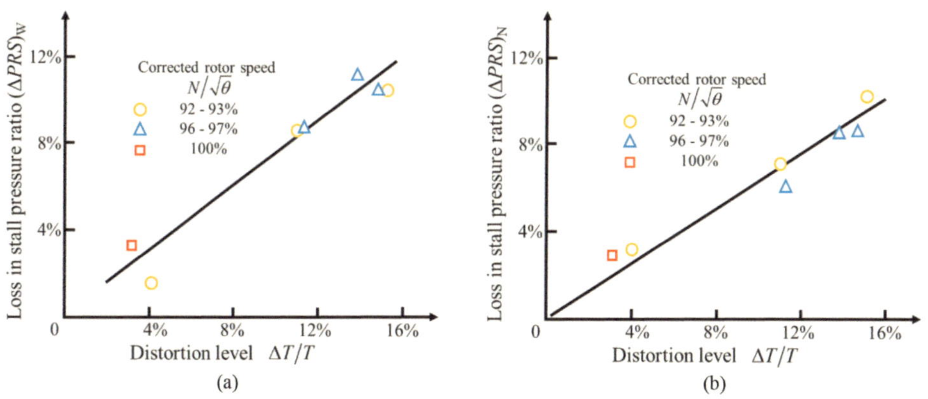


**Figure 12.**
*Loss in the pressure ratio of stall for the 180°-extent temperature distortion [18]. (a) At constant corrected airflow. (b) At a constant corrected speed.*

linear with amplitude $\Delta T/T$ of the temperature distortion, while the ratio of $\Delta(PRS)_W$ to $\Delta T/T$ is approximately 0.75. The loss in stall pressure ratio at constant corrected speed $\Delta(PRS)_N$ also appears to be linear, while the ratio of $\Delta(PRS)_N$ to $\Delta T/T$ is approximately 0.62 [18].

## 5. Effects of combined distortions on the aerodynamic stability of fans

The numerical simulation of binary combined distortions is performed. The stability margin loss of the rotor under different combined distortions is calculated to obtain the relationship between the stability margin loss and each distortion parameter.

### 5.1 Effect of the total pressure and swirl distortion on the aerodynamic stability of fans

The stability margin loss of the rotor increases under the total pressure and swirl distortion when the swirl and temperature distortion increases (see **Figure 13**). When the total pressure distortion increases under the low swirl distortion or decreases under the high swirl distortion, the stability margin loss of the rotor also increases. Binary linear regression is performed on the numerical simulation results, and the regression coefficient is about 91.0%.

The instability mechanism of the total pressure and swirl distortion can be sunnitized as follows. **Figure 14** shows the unwrapped blade-to-blade snapshots of the total pressure and velocity under the maximum stability margin loss at different spans. The speed in the distorted sector is lower than that in the clean sector at 99% of the span. The blockage phenomenon has spread from cascades to the front of cascades, and a protruding low-pressure region with a width of about 10 cascades forms in front of cascades. The cascades after the low-pressure region have obvious stall. Although the rotor is still blocked, the influence degree and range of blockage at 80% of the span are weaker than those at 99% of the span, and the low-pressure region in front of the cascade has been reduced to about 7 cascade widths.

The area of the protruding low-pressure region in front of the cascade has little change at 50% of the span, and blockage is no longer obvious. The number of stalled

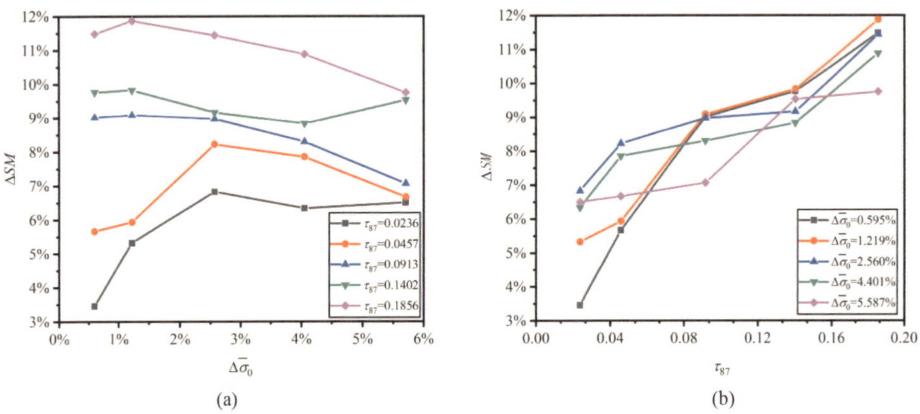


**Figure 13.**
*Stability margin loss under the total pressure and swirl distortion. (a) Total pressure distortion parameters as the abscissa. (b) Swirl distortion parameters as the abscissa.*

(a) Total pressure distribution

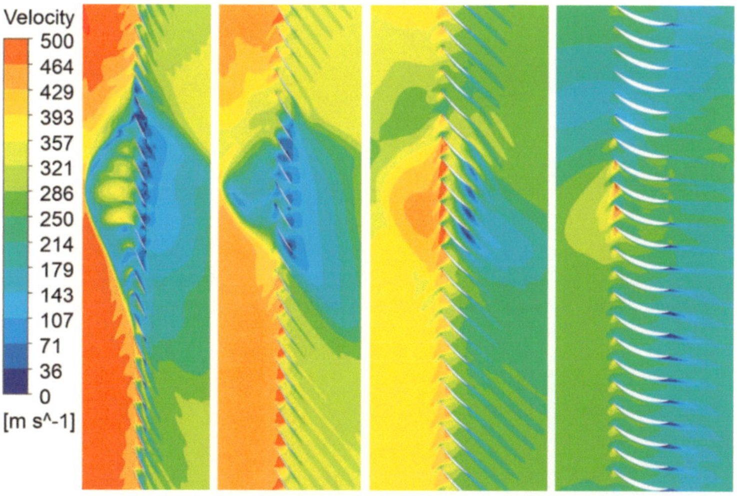


(b) Velocity distribution

**Figure 14.**
*Different spans with a maximum stability margin loss under the total pressure and swirl distortion (from left to right: 99, 80, 50, and 10% of the span). (a) Total pressure distribution. (b) Velocity distribution.*

cascades decreases, and the overall high-speed region appears at the cascade inlet. The protruding low-pressure region in front of the cascade is not obvious at 10% of the span, and flow in the rotor tends to be normal. Therefore, when the rotor is under the total pressure and swirl distortion, stall and blockage appear at the tip first and then gradually spread to the midspan and root of the hub.

## 5.2 Effects of the total temperature and swirl distortion on the aerodynamic stability of fans

The stability margin loss of the rotor increases under the total temperature and swirl distortion when the total temperature distortion decreases or the swirl distortion increases (see **Figure 15**). The binary linear regression is performed on the numerical simulation results, and the regression coefficient is about 93.0%.

The instability mechanism under the total temperature and swirl distortion is generally similar to that under the total pressure and swirl distortion. **Figure 16** shows unwrapped blade-to-blade snapshots of the total pressure and velocity under the maximum stability margin loss at different spans. The blockage phenomenon has spread from the cascades to the front of the cascades at 99% of the span, and a protruding low-pressure region covering the entire circumference has been formed in front of the cascade. Almost all cascades have serious stall phenomena. Although the rotor is still blocked, the influence degree and range of the blockage are weaker at 80%

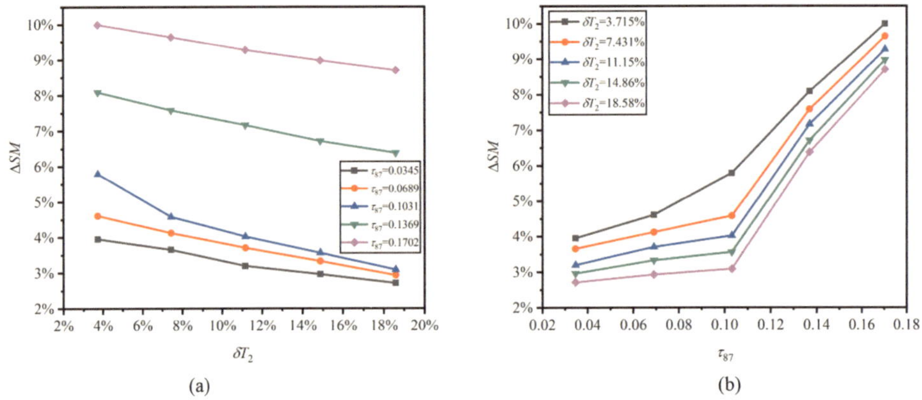


**Figure 15.**
*Stability margin loss under the total temperature and swirl distortion. (a) Total temperature distortion parameters as the abscissa. (b) Swirl distortion parameters as the abscissa.*

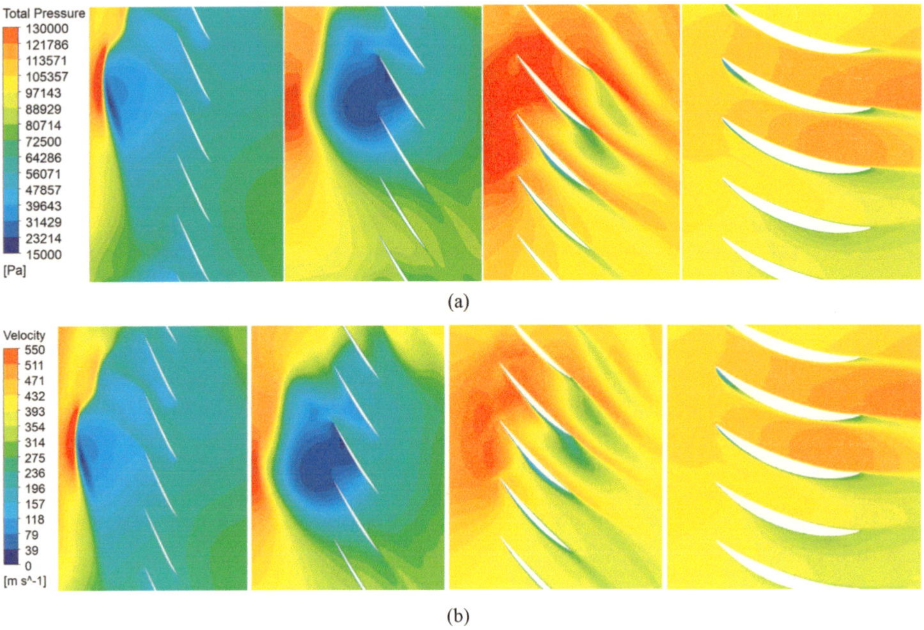


**Figure 16.**
*Different span with a maximum stability margin loss under the total temperature and swirl distortion (from left to right: 99, 80, 50, and 10% of the span). (a) Total pressure distribution. (b) Velocity distribution.*

*DOI: http://dx.doi.org/10.5772/intechopen.109209*

of the span than those at 99% of the span. The low-pressure region in front of the cascade has been reduced to two stall cells, one is big and the other is small.

Flow tends to be normal in the lower blade-height plane, but some cascades still stall. Flow tends to be normal at the span closer to the hub, but some cascades still stall. Therefore, when the rotor is under the total temperature and swirl distortion, stall and blockage appear at the tip first and then gradually spread to the midspan and root of the hub.

### 5.3 Effect of the total pressure and total temperature distortion on the aerodynamic stability of fans

The stability margin loss of the rotor increases under the total pressure and total temperature distortion when the total pressure distortion increases or the total temperature distortion decreases (see **Figure 17**). Binary linear regression is performed on the numerical simulation results, and the regression coefficient is about 84.1%.

The instability mechanism under the total pressure and total temperature distortion is generally similar to that under the total pressure and swirl distortion. **Figure 18** shows unwrapped blade-to-blade snapshots of the total pressure and velocity under the maximum stability margin loss at different spans. When the rotor is under the total pressure and total temperature distortion, stall and blockage appear at the tip first and then gradually spread to the midspan and root of the hub.

### 5.4 Functional relationship between distortion parameters and the aerodynamic stability of fans

First of all, the calculated examples are supplied using a uniform test design to increase the generality, universality, and reliability of the calculation results. Secondly, the mathematical relations between the stability margin loss and distortion parameters are derived and used as the basis for the next calculation and analysis to strengthen the logic of the calculation and analysis (**Figure 19**).

The changes in the velocity triangle before and after the swirl distortion and mathematical approximation show that stability margin loss $\Delta SM$ and swirl distortion

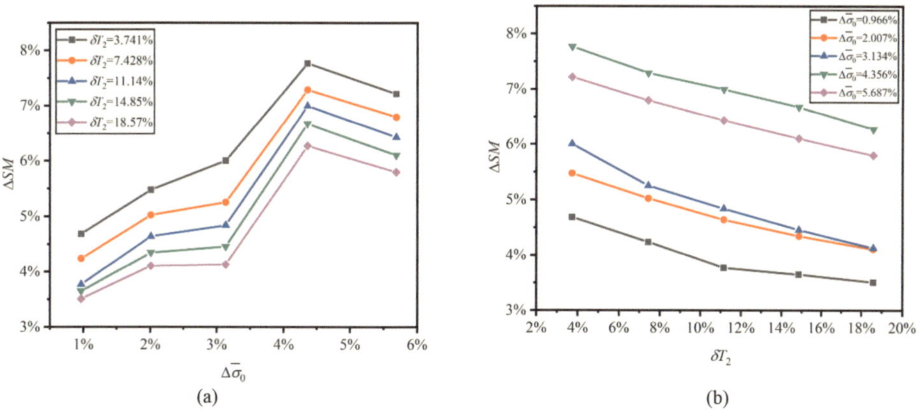


**Figure 17.**
*Stability margin loss under the total pressure and total temperature distortion. (a) Total pressure distortion parameters as the abscissa. (b) Total temperature distortion parameters as the abscissa.*

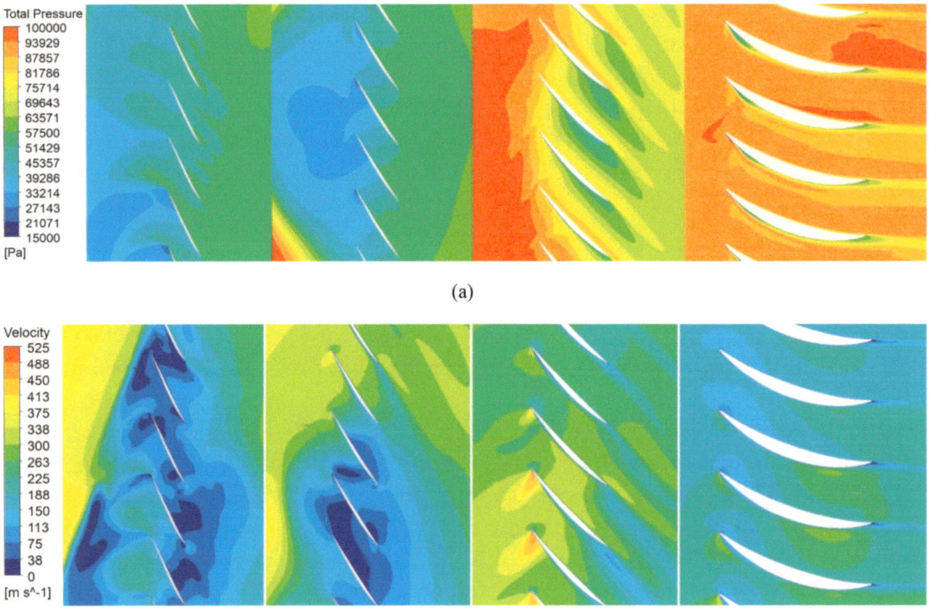


**Figure 18.**
*Maximum stability margin loss under the total pressure and total temperature distortion at different spans (from left to right: 99, 80, 50, and 10% of the span). (a) Total pressure distribution. (b) Velocity distribution.*

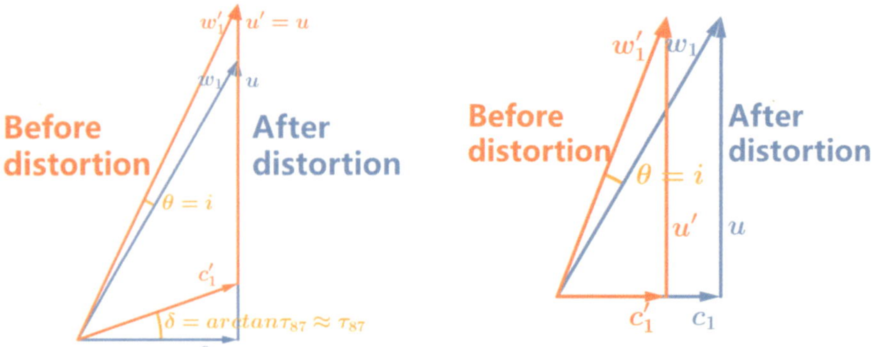


(a) (b)

**Figure 19.**
*Velocity triangles before and after the distortion. (a) Swirl distortion. (b) Total temperature distortion.*

parameter $\overline{\tau}_{87}$ have an approximate quadratic relationship when loss coefficient $\overline{\omega}$ has an approximate quadratic relationship with attack angle $i$ and an approximately linear relationship with stability margin $SM$ (see Eq. (6)).

Similarly, the changes in the velocity triangle before and after the total temperature distortion and mathematical approximation show that stability margin loss $\Delta SM$ and total temperature distortion parameter $\delta T_{2\text{FAV}}$ have an approximately quartic relationship (see Eq. (7)). Besides, stability margin loss $\Delta SM$ and total pressure distortion parameter $\Delta\overline{\sigma}_0$ have an approximately linear relationship (see Eq. (8)).

The approximate error derived from the above mathematical relationship is within the acceptable range through mathematical verification.

$$\Delta SM_\tau \sim \overline{\omega} \sim i^2 \sim \tau_{87}^2 \tag{6}$$

$$\Delta SM_\tau \sim \overline{\omega} \sim i^2 \sim \left(\frac{\Delta T}{T}\right)^4 \tag{7}$$

$$\Delta SM_p \sim \overline{\omega} \sim \frac{\Delta p}{p} \tag{8}$$

According to the numerical simulation results including the ternary combined distortion examples, linear regression fitting is carried out based on the approximate mathematical relationship between the stability margin loss and each distortion parameter derived above. The obtained fitting equation is used as the prediction model of the stability margin loss of fans based on combined distortion parameters (see Eq. (9)).

$$\Delta SM = tr(AX) + B \tag{9}$$

where

$$A = \begin{bmatrix} a_1 & 0 & 0 & 0 \\ a_2 & a_4 & a_6 & a_7 \\ a_3 & a_5 & 0 & 0 \end{bmatrix} \tag{10}$$

$$X = \begin{bmatrix} \Delta\overline{\sigma}_0 & \delta T_2 & \tau_{87} \\ \Delta\overline{\sigma}_0{}^2 & \delta T_2{}^2 & \tau_{87}{}^2 \\ \Delta\overline{\sigma}_0{}^3 & \delta T_2{}^3 & \tau_{87}{}^3 \\ \Delta\overline{\sigma}_0{}^4 & \delta T_2{}^4 & \tau_{87}{}^4 \end{bmatrix} \tag{11}$$

$$B = a_0 \tag{12}$$

**Table 3** shows the parameter values in the prediction model.

Compared with the CFD calculation results, the error of the prediction model is generally within ±3.1%; the average of the absolute value of the error is 0.801%; the median of the absolute error value is 0.610%. To sum up, the prediction results are generally good.

| Parameter No. | Parameter value |
|---|---|
| 0 | 0.058579 |
| 1 | 0.363079 |
| 2 | −0.033317 |
| 3 | −0.124045 |
| 4 | −4.00817 |
| 5 | 2.11732 |
| 6 | 31.9687 |
| 7 | −71.91360 |

**Table 3.**
*Parameter values of the prediction model.*

**Figure 20** shows the prediction results. Scattered points are CFD calculation data, and the 3D counter is the prediction result. The prediction model has a good prediction effect and can be used as a prediction tool for the stability margin loss.

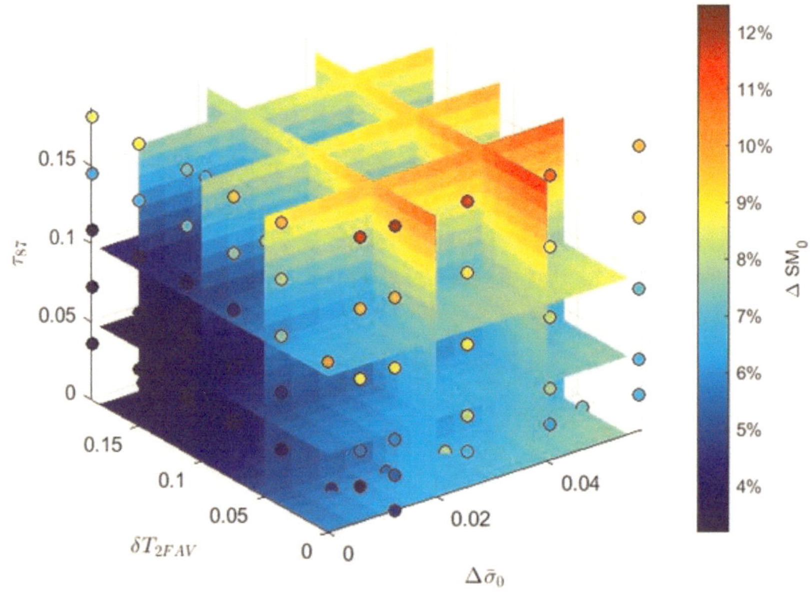


(a)

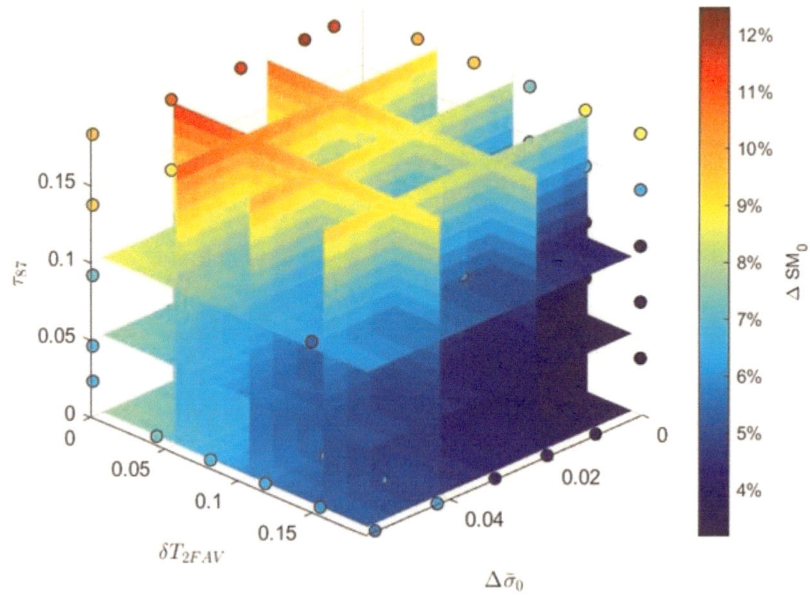


(b)

**Figure 20.**
*Prediction of the stability margin loss under the total pressure, total temperature, and swirl distortion based on the prediction model. (a) View of the low-total-pressure and low-total-temperature distortion. (b) View of the high-total-pressure and high-total-temperature distortion.*

## 6. Conclusions

The 3D CFD numerical simulation was utilized in the work. First, the numerical simulation method was verified to ensure the credibility of the results. Then the influences of combined distortions on the aerodynamic stability of fans were numerically calculated. The following conclusions are obtained:

1. The stability margin loss of the rotor increased under the total pressure and swirl combined distortion when the swirl and temperature distortion increased. When the total pressure distortion increased under the low swirl distortion or decreased under the high swirl distortion, the stability margin loss of the rotor also increased. The instability mechanism showed that stall and blockage appeared at the tip first and then gradually spread to the midspan and root of the hub when the rotor was under total pressure and swirl distortion.

2. The stability margin loss of the rotor increased under the total temperature and swirl distortion when the -distortion decreased or the swirl distortion increased. The instability mechanism under the total temperature and swirl distortion was generally similar to that under the total pressure and swirl distortion.

3. The stability margin loss of the rotor increased under the total pressure and total temperature distortion when the total pressure distortion increased or the total temperature distortion decreased. The instability mechanism under total temperature and total temperature combined distortion was generally similar to that under total pressure and swirl combined distortion.

4. A prediction model of the stability margin loss of fans based on combined distortion parameters was established through linear regression fitting and an approximate mathematical relationship between the stability margin loss and each distortion parameter. The model was used for verification, and the error was within $\pm 3.1\%$ compared with CFD calculation results. The model has good prediction ability and can be used as a prediction tool for stability margin losses.

## Author details

Quanyong Xu[1*], Anping Hou[2] and Honghu Wu[2]

1 Tsinghua University, Beijing, China

2 Beihang University, Beijing, China

*Address all correspondence to: xqybuaa@163.com

## References


[1] Braithwaite WM, Soeder RH. Combined pressure and temperature distortion effects on internal flow of a turbofan engine. Journal of Aircraft. 1980;**17**(7):468-472

[2] Soeder RH, Mehalic CM. Effect of Combined Pressure and Temperature Distortion Orientation on High-Bypass-Ratio Turbofan Engine Stability. Technical report, NASA TM-83771. Cleveland, Ohio, USA: NASA; 1984

[3] Mehalic CM. Effect of spatial inlet temperature and pressure distortion on turbofan engine stability. In: AIAA/ASME/SAE/ASEE 24th Joint Propulsion Conference. Boston, Massachusetts, USA: AIAA; 1988. Paper number AIAA-88-3016

[4] Davis M Jr. Parametric investigation into the combined effects of pressure and temperature distortion on compression system stability. In: AIAA/ASME/SAE/ASEE 27th Joint Propulsion Conference. Sacramento, California, USA: AIAA; 1991. Paper number AIAA-91-1895

[5] Davis MW Jr, Cousins WT. Evaluating complex inlet distortion with a parallel compressor model: Part 2—Applications to complex patterns. In: ASME Turbo Expo 2011. Vancouver, British Columbia, Canada: ASME; 2011. Paper number GT2011-45068

[6] Frohnapfel DJ, Ferrar AM, Bailey JM, O'Brien WF, Lowe KT. Measurements of fan response to inlet total pressure and swirl distortions produced by boundary layer ingesting aircraft configurations. In: 54th AIAA Aerospace Sciences Meeting. San Diego, California, USA: AIAA; 2016. Paper number AIAA-2016-0533

[7] Huang SZ. A numerical study in effect of combined pressure and temperature distortion on engine stability. Gas Turbine Experiment and Research. 2002; **15**(01):28-32 (in Chinese)

[8] Ye W, Hou MJ, Zhou ZW. Numerical study on the effects of pressure and temperature distortion of an engine. Gas Turbine Experiment and Research. 2007; **20**(04):41-44+21 (in Chinese)

[9] Xie YP, Liu YQ, Pan BJ. Aerodynamic calculation method of engine stability under actual inlet condition. Journal of Aerospace Power. 2019;**34**(04):804-812 (in Chinese)

[10] Pearson H, McKenzie AB. Wakes in axial compressors. Journal of the Royal Aeronautical Society. 1959;**63**(583): 415-416

[11] Reid C. The response of axial flow compressors to intake flow distortion. In: Gas Turbine Conference & Products Show. Cleveland, Ohio, USA: ASME; 1969. Paper number 69-GT-29

[12] Mazzawy RS. Multiple segment parallel compressor model for circumferential flow distortion. Journal of Engineering for Gas Turbines Power. 1977;**99**(2):288-296

[13] Davis MW Jr, Cousins WT. Evaluating complex inlet distortion with a parallel compressor model: Part 1—Concepts, theory, extensions, and limitations. In: ASME Turbo Expo 2011. Vancouver, British Columbia, Canada: ASME; 2011. Paper number GT2011-45067

[14] SAE International. Inlet Total-Pressure-Distortion Considerations for Gas-Turbine Engines. Warrendale,

Pennsylvania, USA: Standard No. AIR1419 Revision C; 2017

[15] Peng CY, Ma JJ, Yin JF. Measurement of inlet swirls in flight. Journal of Propulsion Technology. 1994; **04**:8-13 (in Chinese)

[16] Strazisar AJ, Wood JR, Hathaway MD, Suder KL. Laser Anemometer Measurements in a Transonic Axial-Flow Fan Rotor. Technical report, NASA-TP-2879. Cleveland, Ohio, USA: NASA; 1989

[17] Arima T, Sonoda T, Shirotori M, Tamura A, Kikuchi K. A numerical investigation of transonic axial compressor rotor flow using a low-Reynolds-number k–ε turbulence model. Journal of Turbomachinery. 1999;**121**(1): 44-58

[18] Braithwaite WM, Graber EJ Jr, Mehalic CM. The effects of inlet temperature and pressure distortion on turbojet performance. In: AIAA/SAE 9th Propulsion Joint Specialists Conference, Las Vegas. Nevada, USA: NASA TM-X-71431; 1973

Section 2

# Turbomachinery Solutions

Chapter 4

# Electrification for Aero-Engines: A Case Study of Modularization in New Product Development

*Edgar Jeevan Danaraj*


## Abstract

Modularization of hybrid-electric propulsion for commercial aircraft is becoming a reality in air transportation. The main intent of an electric architecture is to produce less carbon emissions and advance towards sustainability in the aeronautics industry. Due to regulatory and customer requirements for new technologies aimed at climate change and pollution, the integration of hybrid electric engine design become more challenging. Conceptual modular and integral product architectures are being compared with conventional and new constructions. A Design Structure Matrix (DSM) model is developed to analyze configuration of sub-component and their relationships through interaction between system elements. The DSM model includes product decomposition and cyclic task interdependencies to understand the extent of modularity in the product life cycle. The traditional turbofan engine architecture will be compared with hybrid electric propulsion engine architecture. The analysis indicates that the electric engine configuration constitutes a shift to a more distributed and less modular architecture. The DSM model reported a 19% increase in density of connectivity between components and 58% decrease in terms of structural complexity. The significance of these changes demonstrates that the more distributed architecture of the fully electric engine architecture requires less effort in system integration than the geared traditional turbofan architecture.


**Keywords:** aircraft, aerospace, engines, electrification, sustainability

## 1. Introduction

The architecture for new and future generation aircraft engines is on the horizon. Building the architecture of a new product is a key design task that affects the steps of product development lifecycle. According to Ulrich, product architecture is the scheme by which the functional components of a product are allocated to physical units [1]. This definition has been widely accepted by many organizations. Comfort, safety, seating capacity, reliability and speed are important factors in today's world of urban air mobility electrification. Rapid technological advancement requires rapid research into stronger lightweight materials and higher battery densities to innovate and grow the product architecture to the next level and embrace the future of commercial flight.

The design of an aircraft engine has been relatively similar for decades, utilizing gas turbine fundamentals since the early twentieth century. Hybrid-electric technology is a crucial means of establishing new standards for aviation sustainability in a variety of applications by maximizing the propulsion system's efficiency. Building blocks of the design structure matrix (DSM) for hybrid-electric engines will be created to compare the architectures between gas turbine and hybrid-electric engines. At the end of the chapter, the matrix will map out and outline product change during technology insertion and solve the question on how a modular architecture could make standardization possible.

## 2. Gas turbine engines

The typical aero-engine is designed in sectional-modular architecture type that can be assembled and disassembled, or even interchanged with modules from another compatible engine, without the need for specific retesting or balancing. This unique advantage supports production maintenance and reduces process downtime for the operator. The target to reduce emissions caused by conventional gas turbine engines in realization of the growing air transport demand drives research for new product architectures.

Previous work lacks the understanding on how interactions between components behave in varying modular platforms. Hence, it is important to know the dependency of each individual component (dependent, parallel, or coupled) in relation to the coordinated activities in the development lifecycle. **Figures 1-5** shows various gas turbine engine configurations and interconnected components.

## 3. Hybrid and electric engines

Hybrid aircraft engines will combine the advantages of a turbine engine, like the high energy density of jet fuel, over those of an electric engine, like lower noise levels and less upkeep. Existing battery powered electric propulsion systems in light UAVs and quadcopters provide environmentally friendly concepts with significantly reduced carbon footprint. Air traffic management, adverse weather conditions and battery capacity required for long range travel, energy-to-weight ratio and cooling efficiency are some constraints often debated during technological development of hybrid design concepts.

**Figure 1.**
*Modular engine components in a typical engine architecture [2].*

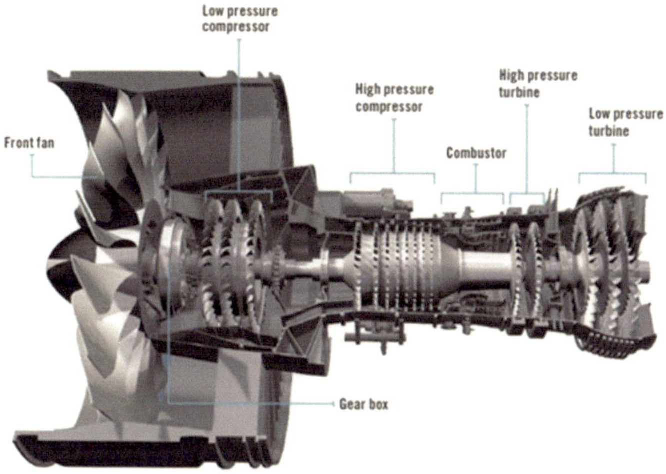


**Figure 2.**
*Pratt and Whitney geared turbofan [3].*

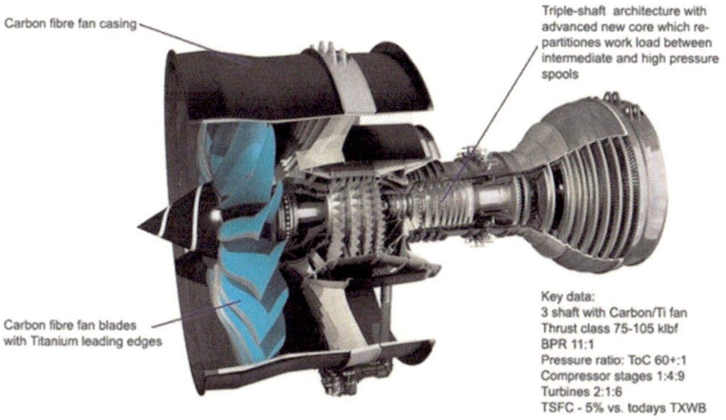


**Figure 3.**
*Rolls-Royce ultrafan engine [4].*

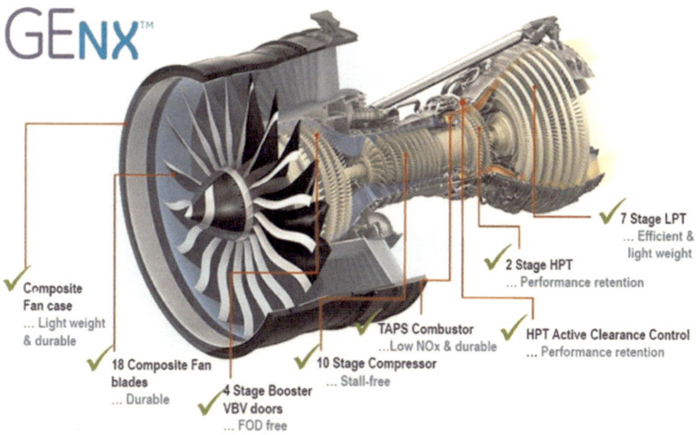


**Figure 4.**
*General electric NX engine [5].*

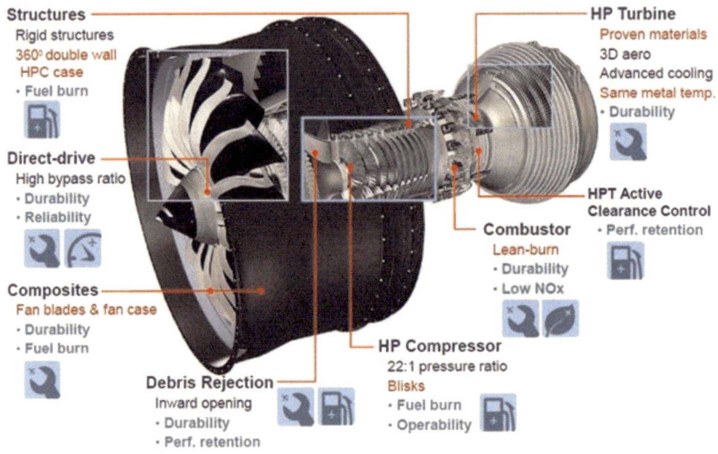


**Figure 5.**
*CFM leap engine [6].*

Gas turbine engines have mechanically turned rotors, propellers, or fans with fuel-burning engines. However, a distributed electric propulsion architecture, in which several electric motors can be tilted or turned off for both vertical take-off and horizontal flight, is used in a lot of recent designs. Anywhere on an aircraft, a turbo-generator can supply electric power to multiple electric motors.

### 3.1 Boeing and Aurora

The all-electric autonomous passenger air vehicle Aurora Flight Sciences, a subsidiary of aerospace giant Boeing, flew for the first time in a test flight. The prototype, according to Boeing, is powered by an electric propulsion system and designed for fully autonomous flight from take-off to landing, with a range of up to 50 miles. Its airframe coordinating the drive and wing frameworks to realize proficient lift and forward flight. Boeing's futuristic NeXt division, which also oversees its heavy-duty drone prototype that can lift a 500-pound payload, carried out the test. Aurora Flight Sciences have collaborated with Uber to create a arrange of "flying taxis" that work on "Uber Air" item, which it plans to dispatch in 2023.

### 3.2 Airbus, Rolls-Royce, and Siemens

In 2017, a 100-seat regional jet with a hybrid electric drive for a turbine was developed by Siemens, Airbus, and Rolls Royce. The E-Fan X hybrid-electric innovation demonstrator was temporarily flown on a BAe 146 flying testbed, with one of the aircraft's four gas turbine motors supplanted by a 2-megawatt electric engine. Arrangements were made to replace a second gas turbine with an electric engine once framework development had been demonstrated.

The compact 2.5 MW (3400 hp) generator was run in 2019 and first flight was scheduled for 2021. The AE2100 turboprop was from a Saab 2000 feeding the battery pack and a Siemens SP2000 electric motor (with a 10 kW/kg power-to-weight ratio) replacing one Honeywell LF507 engine with a Rolls-Royce AE 3007 fan through a 3000 volts AC/DC distribution. However, the program was canceled in 2020 due to the COVID-19 pandemic.

### 3.3 Rolls-Royce

Distributed propulsion, in which a single turbine turns multiple propellers on an aircraft, is made possible by hybridization. Rolls-Royce and APUS, a company that specializes in aviation engineering, and the Brandenburg University of Technology (BTU) are creating a hybrid-electric flight demonstration aircraft. The M250, the engine of choice for hybridization served as the foundation for the eVTOL concept developed by Rolls-Royce. The M250 can power a four- to five-passenger vehicle that can travel at 250 mph over a range of at least 500 miles by adding an electrical generator to the system. The gas turbine produces electricity between 300 and 400 kW, and a battery system can supply an additional 300 to 400 KW for hovering. The aircraft can also perform a standard take-off and landing. All-electric VTOLs for short-range missions will emerge in the longer term as battery technology advances, but hybrids will likely continue to serve longer-range missions (**Figure 6**).

### 3.4 General electric and XTI

GE Aviation's business and general aviation unit's leader in hybrid-electric pursuits. The TriFan 600 will be powered by a new hybrid-electric propulsion system developed by GE Aviation and XTI Aircraft Company. This contrasts with conventional turboprop engines, which require a distinct turbine for each propeller. It makes it possible for a greater number of hybrid aircraft to utilize a greater number of propulsion sources, allowing aircraft designers to re-evaluate even the most fundamental aspects of aircraft design. The current battery technology does not allow for sufficient energy density to make a long-distance electric aircraft feasible.

The TriFan 600's Catalyst will enable it to travel at a much higher altitude (30,000 feet) and at a faster rate than any electric aircraft currently on the market. The Catalyst's power will also make it possible to transport a much larger

**Figure 6.**
*Rolls-Royce M50 engine for the APUS i-5 plane [7].*

**Figure 7.**
*General electric and XTI TriFan 600 [8].*

payload while still being able to take-off and land vertically. GE engineers have been able to reduce the typical 800 components that would have been made using conventional methods down to just a dozen or so printed parts using 3D printing. They were able to use less fuel and reduce the Catalyst's weight by 5% with this strategy. The TriFan's engine will have approximately 1400 horsepower, or 1 megawatt of power (**Figure 7**).

### 3.5 Honeywell

Honeywell's hybrid-electric propulsion includes a turbogenerator that produces 400 kilowatts by combining two miniaturized generators with the robust, flight-tested HTS900 engine. The turbogenerator system can supply motors or high-capacity batteries with conventional or bio-derived jet fuel. In addition, Honeywell introduced the first 1 MW generator for the aerospace industry, which had the same power density as our 200 kVA generator. The 1 MW generator has a weight of 280 lb. The HGT1700 auxiliary power unit, which is currently used on every Airbus A350 XWB, will be combined with this generator to create a turbogenerator that is 2.5 times more powerful than the version that the company unveiled in 2019 (**Figure 8**).

### 3.6 Collins aerospace and Pratt & Whitney Canada

A P&WC engine and a Collins electric motor of 1 megawatt will be combined in a parallel hybrid configuration with a 50/50 power split on one side of an experimental De Havilland Dash 8–100 aircraft. The motor by Collins has a high-power density and efficiency. The program is aiming for a 30% increase in fuel efficiency and correspondingly lower CO2 emissions. The fuel-burning engine will be able to be optimized

**Figure 8.**
*Honeywell 1-megawatt turbogenerator [9].*

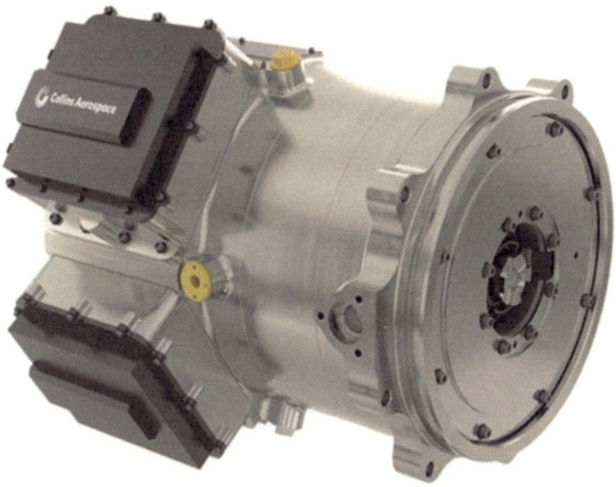


**Figure 9.**
*Collins aerospace 1-Megawatt electric motor [10].*

for cruise efficiency because the electric motor will provide additional power during take-off and climb. Additionally, the new technology will be constructed to operate entirely on Sustainable Aviation Fuel (SAF) (**Figure 9**).

### 3.7 Lilium

Ducted Electric Vectored Thrust (DEVT) is a proprietary technology from Lilium Jet, which incorporates electric jet engines into the wing flaps for thrust vectoring. This provides advantages in payload, aerodynamic efficiency, and a lower noise profile. Each of the 36 individually controllable flaps in the propulsion system has a ducted electric fan and serves as a lifting and control surface. On the canard and main wing, the 36 ducted fans are embedded in a 1:2 ratio. The ducted fans are incorporated into the wings, reducing weight, and minimizing aerodynamic drag loss by eliminating the need for separate nacelles.

**Figure 10.**
*Lilium electric jet engine [11].*

Because the discharge capacity of the batteries did not provide sufficient power at lower levels of the SOC, the battery used permitted a minimum SOC of approximately 30–40%. Modern cell technology employs more advanced anode materials, such as silicon, to increase the battery cell's discharge capacity and, as a result, significantly improves power provision at low SOC. A minimum SOC of 10–15% is enabled for the Lilium Jet, which has seven seats.

Over the years, the performance of the development of battery cells has rapidly improved, and they are now available with energy densities of more than 300 Wh/kg and power densities of more than 3 kW/kg. This helps with the design of a battery system with more than 300 kWh of total stored energy, to achieve a maximum physical range of 250 or more kilometers, including reserves (**Figure 10**).

## 4. Advantages of electric engines

### 4.1 Noise

An electric engine is more silent than a motor that combusts fuel. It must drive a rotor, propeller or fan which produces noise during take-off. But electric engines will be quieter when taxiing and cruising. Moreover, distributed propulsion systems with multiple smaller, quieter rotors or fans are also made possible by electric motors.

### 4.2 Efficiency

Another benefit is efficiency. Compared to today's large turbofans, which are 55% efficient, and small turboprops, which are 35%, electric drivetrains can be more than 90% efficient. One of the reasons why the electrification of propulsion is beginning with the modification of regional aircraft powered is because of the disparity in efficiency between large and small turbines.

## 5. Disadvantages of electric engines

### 5.1 Battery

The typical lithium-ion battery is utilized in electric aircrafts can be associated with hazards such as short-circuit and chemical leakage. There are battery chemistries that perform significantly better than current lithium-ion but are not available in the market. The automotive industry is utilizing hydrogen fuel cells and other forms of energy storage. The flow battery, which NASA is exploring for the Aquifer project, is one illustration of novel energy storage within the early stages of development.

### 5.2 Weight

Super-lightweight carbon-fiber construction methods can be used to make airplanes lighter. A better design can be used to reduce required power. For the longer endurance mission requirements of the largest aviation market segments, a hybrid solution that incorporates a small jet-fuel-powered auxiliary power unit can be utilized. While avoiding the expense and weight of an all-electric battery-powered aircraft, this solution offers improved efficiency at a lower cost. The advantages of electric propulsion will be augmented by jet fuel's high energy density. Additionally, lithium-ion batteries, which are lighter, can be utilized.

### 5.3 Range

Utilizing electric planes will not be feasible due to their limited range, despite the environmental and financial benefits of not using fossil fuels and the emissions that come with it. Electric planes have the same range as cars on the road, which ranges from 160 to 400 kilometers.

### 5.4 Thermal runaway

The electrical drive for an aircraft engine must be thermally stable and reliable to pass airworthiness regulations and be deemed safe for flight. Various electrical concepts can be used to design a fully electric aircraft. With regards to the existing electric engine architecture, the motor does not have a robust high torque and insufficient power density under heavy payloads.

## 6. Electric motor concepts

**Table 1** characterizes various electric motor concepts. The permanent magnet synchronous machine is shown as the most suitable design for an electrical architecture (**Figure 11**). However, to qualify and implement this concept, many years of testing and data is required to prove its feasibility (**Figure 12**).

## 7. Significance of product design

The foundation of product design and development stretches across product architecture, new technology and customer requirements. Product architecture for

| Key Characteristic | ESM | IM | SRM | PMSM |
|---|---|---|---|---|
| Rotor losses | — | o | o | ++ |
| Stator losses | ++ | o | o | o |
| Windage Losses | — | o | — | ++ |
| Rotor thermal limitations | o | + | ++ | o |
| Cooling options | — | o | o | ++ |
| Rotor mechanical limitations | — | o | + | ++ |
| Torque-to-inertia ratio | o | o | o | ++ |
| Compatibility with bearings | — | o | o | ++ |
| High-speed capability | — | o | + | ++ |
| Short-circuit behavior | — | ++ | ++ | — |
| Machine complexity | o | + | ++ | + |
| Current density | — | + | + | + |
| Power density | — | + | + | + |

*Excited Synchronous Machine (ESM); Induction Machine (IM); Switched Reluctance Machine (SRM); Permanent Magnet Synchronous Machine (PMSM); – unfavorable; – disadvantageous; o neutral; + beneficial; ++ very beneficial.*

**Table 1.**
*Key characteristics of various electric motor concepts [12].*

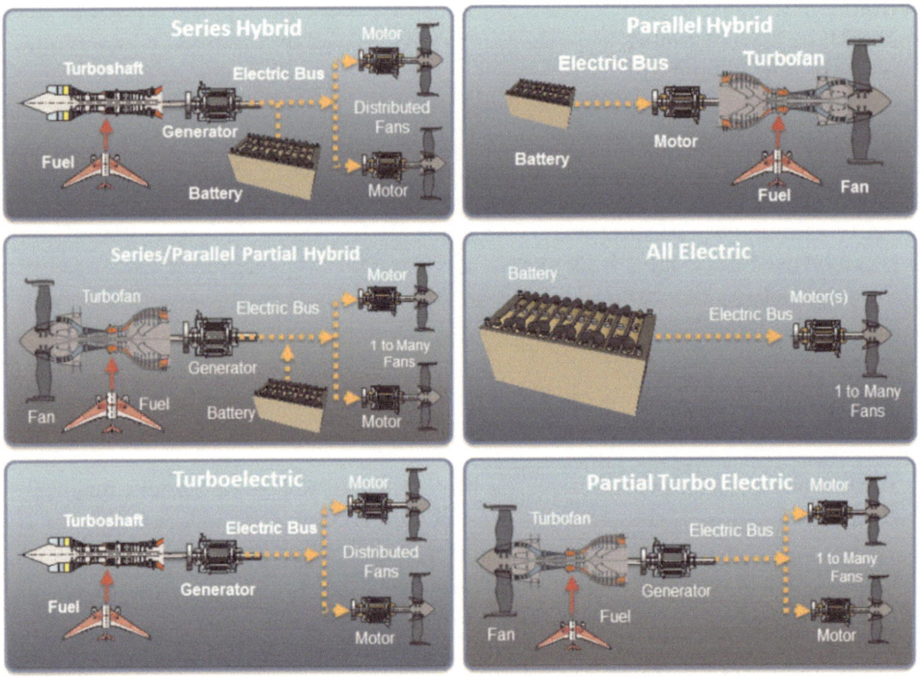


**Figure 11.**
*Schematic of electric propulsion system architectures [13].*

**Figure 12.**
*Substituting an electric drive system in a gas turbine engine (hybrid).*

aircraft propulsion design relies on modularity and evolving technologies. The main scope discussed is the future of successful aircraft engine design for the mass transportation market. For strategic technology insertion, the effects of introducing hybrid electric propulsion systems in modern day passenger aircraft will be analyzed.

Different engine modularity and its limitations will be briefly discussed and how it will affect the product development process. Scope and approach give a breakdown on the product lifecycle and when to insert new technology for a successful outcome.

## 8. The product development process

In product development process, the modular and integral approach is defined to understand in detail the different characteristics involved in reconfiguration for a manufacturing system. Strategic technology insertion and challenges in manufacturing will also be briefly discussed. Product architecture is determined in the early stages of process development. Functional elements are arranged into specific blocks to breakdown their functions (what they do) and how they are interfaced to the rest of the product (**Figure 13**).

**Modular approach**

i. Functional component elements are mapped

ii. Protocols and interface standards are defined

iii. Component designs are in parallel

iv. Component testing can be done independently

v. Checking for unanticipated coupling and interactions is focused

vi. Required performance changes are localized to a few components

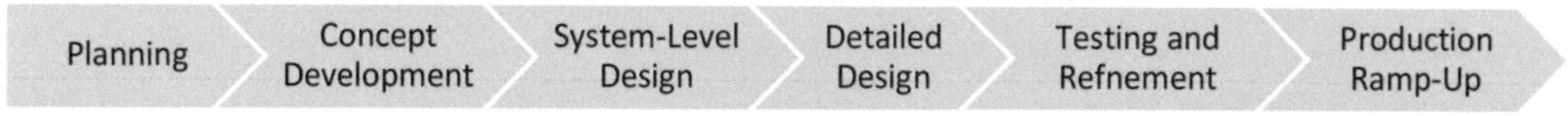


**Figure 13.**
*The product development process.*

### Integral approach

i. System-Level performance targets are emphasized

ii. Product is divided into few integrated sub-systems

iii. Component tests must be done simultaneously

iv. Effort focused on tuning the overall system

v. Required performance changes propagate to many components

In choosing either one or both two architectural approaches, we need to define the technical working principles and desired variety of features, including setting performance targets. Two indicative definitions are “the ability to repeatedly change and rearrange the components of a system in a cost-effective way” [13] and “the ability of a function of a manufacturing unit to be simply altered in a timely and cost- effective manner” [14].

The reconfigurability of a manufacturing system can be further understood in terms of certain characteristics it exhibits. Modularity is the extent to which all system components; both software and hardware are modular. Integrability is the ability in which systems and components are combined with the introduction of new technology.

### 8.1 Definition of modularity

The term “modularity” has been widely used to suggest decoupling of construction structures, such that the more decoupled the structures of a product or system, the more modular that product or system will tend to be [15].

$$\text{Component modularity} = \frac{\text{Actual component disconnectivity}}{\text{Maximum likely component disconnectivity.}} \tag{1}$$

Modularity is the extent in which a design system may be split into segments and merged back again. Modularization is intended to effectively redistribute total complexity throughout the system by clustering elements into chunks. A complex modular architecture with multiple modules can result in a dense architecture but should still be advantageous if system decomposability is of major importance. This primarily relates to the effectiveness of using reductionist approaches. Increasing a product’s modularity enables these strategies whereas higher complexity makes reductionism less effective.

Modularity is mathematically defined as the extent at which two architectures correspond, on the contrary of intersection. The calculation of correspondence, the Correspondence Ratio (CR), is stated as [16]:

$$CR = \frac{|Vi(x) \cap Vj(x)|}{|Vi(x) \cap Vj(x)|} \tag{2}$$

where |X| points out the number of elements (cardinality) of group X. CR will be near to 1 if the correlation between the two modules is high, vice versa. This is a good way to ascertain module by module foundation, however, does not give good comparison between modules of different designs. A more reliable consideration of module correspondence for an entire product is the average CR for all modules in the product, CRoverall [16]:

$$\text{CRoverall} = \frac{\Sigma CRi}{\#Modules} \tag{3}$$

Likewise, CRoverall = 0 signifies that there is no correspondence between perspectives. CRoverall = 1 when the individual module CRs approaches 1. To convey the second attribute of modularity, reducing minor interactions, a Cluster Independence (CI) is defined [16].

$$\text{Modularity} = (\text{CRoverall}) \times (\text{CI}) \tag{4}$$

### 8.2 Definition of modularity

Apart from new product introduction (NPI) and managing design changes, strategic technology insertion is one of the important factors in life-cycle management. The three sections where the benefits of a specific technology begin to stagnate are:

i. Technology introduction

ii. Technology improvement

iii. Technology maintenance

In new product development, the critical success factors in descending order of success likelihood are Strong Market Orientation (high), Early Planning and Specification (medium), and Technical and Marketing Excellence (low).

## 9. Challenges in manufacturing

The main challenges in today's Manufacturing world are customer satisfaction and new technology. Customers have higher expectation for product quality (performance) and on-time delivery at lower cost and shorter time to manufacture. This new paradigm of product development success "to provide the greatest value at the lowest possible cost and shortest time" is hard to achieve and impossible in some situations. The evolving landscape of new technology, digital manufacturing innovation and smart product design influence the strategic decision-making process for both customer and supplier. The lifespan of a new product is shorter nowadays, creating a demand for modularity and commonality weaved together to maximize value and use.

At such an increasing rate of rapid change, the product configuration must be managed strategically so that the organization or business will stand to gain profit and minimize losses. Because of this, most products are designed and built in a fashion that allows it to be remanufactured or reconfigured for reuse, and in certain cases, the quality or performance of the product may be compromised. This is a big concern, especially for reputable companies who have brand reputation at stake. Management finds it difficult to focus on time instead of cost. Period of profitability, time to design, late product launch and competition from rival companies add to the complexity of challenges, where the voice of the customer is always focused on center stage.

## 10. Design structure matrix (DSM)

This section assesses the methodology of the design structure matrix in terms of modularity and structural complexity. The design structure matrix for three product families to be developed will be used for comparison of the different architectures. The three models are:

i. Old architecture (normal gas turbine engine without a gearbox system)

ii. Existing architecture (geared turbofan engine or hybrid metallic fan blade)

iii. New architecture (hybrid or fully electric jet engine)

The main advantage in mapping out these architectures is to provide a platform to overlay modules and individual parts on the design structure. This ultimately may explain how the different architectures contribute to the product development process through interactions.

The design structure matrix is conceptualized with the modules to provide the means to approach complexity and connectivity between the teams accountable for the design and fabrication of the engine parts. The method in which the design structure matrix was generated is considered in system level decomposition, applicable to all platforms equally. Engine parts selected were dependent upon its functional representation in the system.

In the design structure of the matrix, multiple parts and stages per module were simplified to one. Repetitive units in the system do not add design complexity and hence are not included. Having identified parts significant to the system architecture, relationship of part-to-part components were populated using value mapping and encoding scheme. In **Table 2**, the outlined blocks indicate the 5 modules found by applying the modularity metric. In **Table 3**, the outlined blocks indicate the 16 modules found by applying the modularity metric. In **Table 4**, the outlined blocks indicate the 5 modules found by applying the modularity metric.

## 11. Product architecture comparison and structural complexity

The three architectures discussed display inherent differences in terms of density and connectivity. This is probably due to the existing architecture having a higher integration of the geared components, packed inside small-scale housing. As a result, structural complexity rises, and can be validated by computed metrics. The structural complexity and modularity metrics were calculated together with interconnections to show variations in their architectural properties. The formula for structural complexity is defined in the following as (**Figure 14**):

Structural complexity distinguishes the system architecture (scheme of interactions), determined by the form. Complexity of the individual components is composed by the total number of $\alpha$'s, to the addition with contribution of complexity due to the number of component interactions and their order. The second term is the

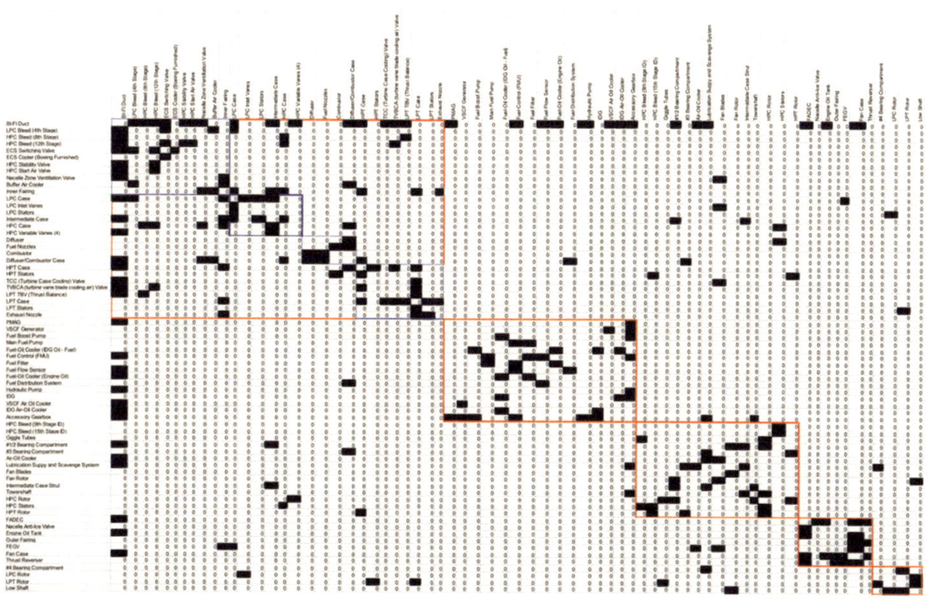

**Table 2.**
*DSM of old architecture [17].*

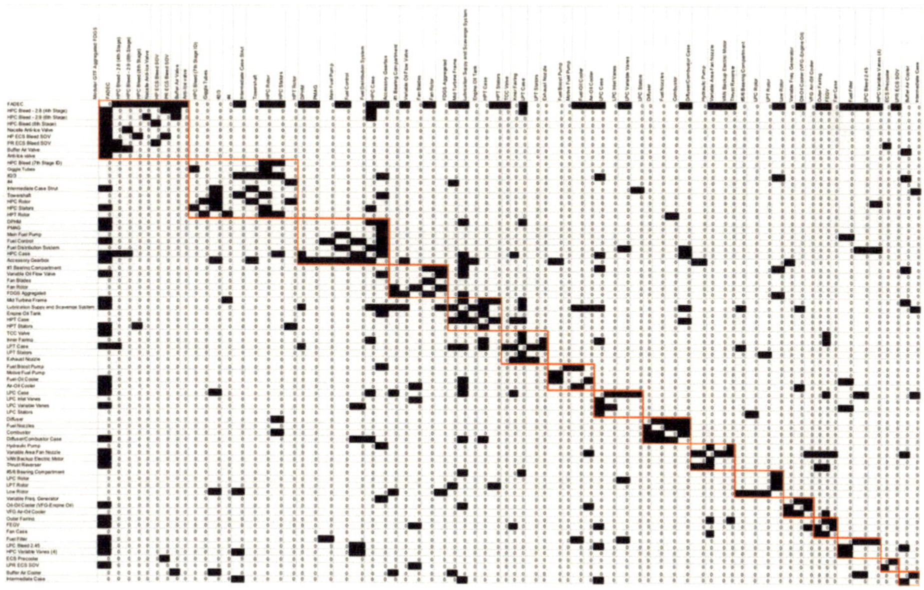

**Table 3.**
*DSM of existing architecture [17].*

increased outcome of the total number of interfaces and graph energy. Component complexities are evaluated and exhibit the internal complexity of individual components in the structure (**Figure 15**) (**Table 5**).

| | Fan Blades | Fan Rotor | Fan Case | Thrust Reverser | Bearing | Low Shaft | ECU | BCU | MCU | Electric motor | Inverter/converter | Drive & Shield | Switching electronics | Sensors | Brushes & Terminals | Electric cable | Terminal box | Li Battery | Alternator | Charging unit |
|---|---|---|---|---|---|---|---|---|---|---|---|---|---|---|---|---|---|---|---|---|
| Fan Blades | 0 | ■ | 0 | 0 | 0 | 0 | 0 | 0 | 0 | 0 | 0 | 0 | 0 | 0 | 0 | 0 | 0 | 0 | 0 | 0 |
| Fan Rotor | ■ | 0 | ■ | 0 | 0 | ■ | 0 | 0 | 0 | ■ | 0 | 0 | 0 | 0 | 0 | 0 | 0 | 0 | 0 | 0 |
| Fan Case | ■ | ■ | 0 | 0 | 0 | 0 | 0 | 0 | 0 | 0 | 0 | 0 | 0 | 0 | 0 | 0 | 0 | 0 | 0 | 0 |
| Thrust Reverser | 0 | 0 | 0 | 0 | ■ | ■ | 0 | 0 | 0 | 0 | 0 | ■ | 0 | ■ | 0 | 0 | 0 | 0 | 0 | 0 |
| Bearing | 0 | ■ | 0 | 0 | 0 | ■ | 0 | 0 | 0 | 0 | 0 | 0 | 0 | 0 | ■ | 0 | 0 | 0 | 0 | 0 |
| Low Shaft | 0 | 0 | 0 | ■ | ■ | 0 | 0 | 0 | 0 | 0 | 0 | 0 | 0 | 0 | 0 | 0 | 0 | 0 | 0 | 0 |
| ECU | 0 | 0 | 0 | 0 | 0 | 0 | 0 | ■ | ■ | 0 | 0 | 0 | 0 | 0 | 0 | 0 | 0 | 0 | 0 | 0 |
| BCU | 0 | 0 | 0 | 0 | 0 | 0 | ■ | 0 | ■ | 0 | 0 | 0 | 0 | ■ | 0 | 0 | 0 | 0 | 0 | 0 |
| MCU | 0 | 0 | 0 | 0 | 0 | 0 | ■ | ■ | 0 | 0 | 0 | 0 | 0 | 0 | 0 | 0 | 0 | 0 | 0 | 0 |
| Electric motor | 0 | 0 | 0 | ■ | 0 | 0 | 0 | 0 | 0 | 0 | ■ | ■ | ■ | 0 | ■ | 0 | 0 | ■ | 0 | 0 |
| Inverter/converter | 0 | 0 | 0 | 0 | 0 | 0 | 0 | 0 | 0 | ■ | 0 | 0 | ■ | 0 | 0 | 0 | 0 | 0 | 0 | 0 |
| Drive & Shield | 0 | ■ | 0 | 0 | 0 | ■ | 0 | 0 | 0 | ■ | ■ | 0 | 0 | 0 | 0 | 0 | 0 | 0 | 0 | 0 |
| Switching electronics | 0 | 0 | 0 | 0 | 0 | 0 | 0 | 0 | 0 | ■ | ■ | ■ | 0 | 0 | 0 | 0 | 0 | 0 | 0 | 0 |
| Sensors | 0 | 0 | 0 | 0 | 0 | 0 | 0 | 0 | 0 | 0 | 0 | 0 | 0 | 0 | 0 | ■ | 0 | 0 | 0 | 0 |
| Brushes & Terminals | 0 | 0 | 0 | 0 | 0 | 0 | 0 | 0 | 0 | 0 | 0 | 0 | 0 | ■ | 0 | ■ | 0 | 0 | 0 | ■ |
| Electric cable | 0 | 0 | 0 | 0 | 0 | 0 | 0 | 0 | 0 | 0 | 0 | 0 | 0 | ■ | ■ | 0 | 0 | ■ | 0 | 0 |
| Terminal box | 0 | 0 | 0 | 0 | 0 | 0 | 0 | 0 | 0 | 0 | 0 | 0 | 0 | ■ | ■ | ■ | 0 | 0 | 0 | 0 |
| Li Battery | 0 | 0 | 0 | 0 | 0 | 0 | 0 | 0 | 0 | 0 | 0 | 0 | 0 | 0 | 0 | 0 | ■ | 0 | 0 | ■ |
| Alternator | 0 | 0 | 0 | 0 | 0 | 0 | 0 | 0 | 0 | 0 | 0 | 0 | 0 | 0 | 0 | 0 | ■ | ■ | 0 | 0 |
| Charging unit | 0 | 0 | 0 | 0 | 0 | 0 | 0 | 0 | 0 | 0 | 0 | 0 | 0 | 0 | 0 | 0 | 0 | ■ | ■ | 0 |

**Table 4.**
*DSM of new architecture.*

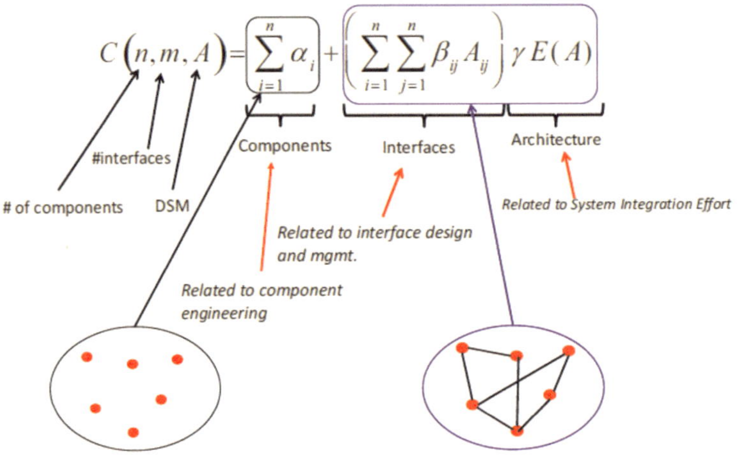


**Figure 14.**
*Structural complexity equation [18].*

The DSM for the new architecture shows significantly more connectivity in all areas measured, with a 19% increase in connection density of the DSM. The individual connection types all decreased in number, indicating a less inter-connected

*DOI: http://dx.doi.org/10.5772/intechopen.109006*

| | **Old Architecture** *Spooled*  | **Existing Architecture** *Geared*  | **New Architecture** *Electric*  | **Change (%)** |
|---|---|---|---|---|
| No. of Components | 71 | 75 | 58 | −23% |
| Connection Density | 5.75% | 6.9% | 8.2% | 19% |
| Total no. of Connections | 271 | 363 | 195 | −46% |
| Mechanical | 242 | 328 | 183 | −44% |
| Information | 49 | 50 | 65 | 30% |
| Energy | 60 | 62 | 40 | −35% |
| Flow | 89 | 107 | 67 | −37% |
| Graph Energy, E(A) | 104.4 | 123.3 | 136.8 | 11% |
| Modularity Index (Q) | 0.43 (5 modules) | 0.35 (16 modules) | 0.62 (5 modules) | 77% |
| Structural Complexity | 550 | 770 | 320 | −58% |

**Table 5.**
*Comparison of all three architectures.*

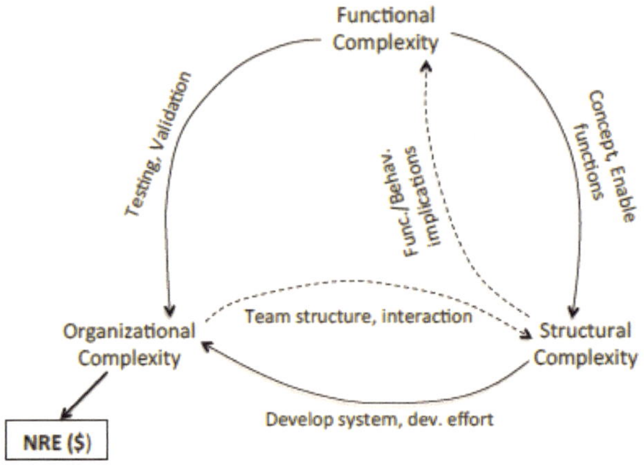


**Figure 15.**
*Main dimensions of complexity in product development [19].*

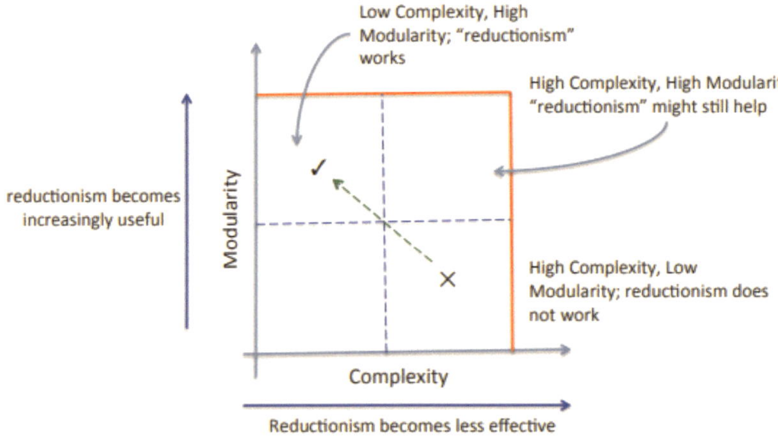


**Figure 16.**
*Complexity-modularity: Impact and effectiveness [20].*

architecture. The largest increase, 30%, is found for the information. This lower level of interconnectivity deduces that the electric engine is significantly more "modular", and this is evident in the high modularity (Q) index.

The increase in graph energy E(A), show that the electric system is more distributed than the geared architecture in the table. The modularity analysis demonstrated for the different engine architectures (total connectivity) of the matrix reveal many inherent modules. The electric architecture is significantly less complex.

**Figure 15** represents ideal quadrants and its effective zones (low complexity, high modularity) where strategies work well using decomposition techniques to tackle system design and development (**Figure 16**).

## 12. Conclusion

A high level of modularity in product design is essential if interactions between chunks are well-defined and implemented by individual functional elements. Having a modular architecture enables a design swap chunks independently without affecting the overall system in terms of functionality. The design structure matrix (DSM) developed analyses the various product architectures, comparing existing with new technology. The data suggest that with an increase in connectivity across the components, a balanced distribution in architecture is reached. The decrease in architectural complexity allows a significant increase in low-cost manufacturability.

The correlation between the DSM (architecture) and functional groups can be summarized into two hypotheses for future research. Hypothesis 1: Components with increasing modularity levels are more likely to be more effective in reductionism. Hypothesis 2: Components with increasing complexity levels are more likely to be less effective in reductionism.

From a strategic technology insertion point of view, data such as architectural complexity and integration cost may be useful in business decisions in the development lifecycle. Electrification is gaining significance in futuristic architecture that is bound to disrupt the aviation sector to make the product development process more coherent.

## Acknowledgements

A research paper on Product Design and Development, School of Mechanical & Aerospace Engineering, Nanyang Technological University.

## Conflict of interest

The authors declare no conflict of interest.

## Author details

Edgar Jeevan Danaraj
Nanyang Technological University, Singapore

*Address all correspondence to: edgarjee001@e.ntu.edu.sg

## References


[1] Ulrich K. The role of product architecture in the manufacturing firm. Research Policy. 1995;**24**(3):419-440

[2] SP's aviation. Issue: 03-2009, Honeywell – The modular, reliable, high-performance solution ready for Jaguar. Available from: http://www.sps-aviation.com/story/?id=236

[3] Read B. Royal aeronautical society, pioneers of power. 26 May 2017. Available from: https://www.aerosociety.com/news/pioneers-of-power/

[4] Sergiu I, Alix F, Timothy M, Joseph D. Transcontinental Windowless Business Jet - Aerospace Vehicle Design - Scientific Figure on ResearchGate. Available from: https://www.researchgate.net/figure/Features-of-the-engine-that-will-be-used-the-Rolls-Royce-Ultrafan-successor-of-the_fig3_317090545 [Accessed: 14 Dec 2022]

[5] Leeham News and Analysis. PW GTF-CFM LEAP market share. Available from: https://leehamnews.com/2012/02/15/pw-gtf-cfm-leap-market-share/

[6] CFM International. The leap engine. Available from: https://www.cfmaeroengines.com/engines/

[7] Derby RJ. Rolls-Royce takes 'significant step' towards hybrid-electric powered aircraft goal. 11 Nov 2019. Available from: https://www.business-live.co.uk/technology/rolls-royce-takes-significant-step-17215225

[8] Haines TB. XTI TRIFAN 600 draws the curious at NBAA. 11 October 2017. Available from: https://www.aopa.org/news-and-media/all-news/2017/october/11/xti-trifan-600-draws-the-curious-at-nbaa

[9] Honeywell Aerospace. Honeywell 1-megawatt (MW) turbogenerator. Available from: https://aerospace.honeywell.com/us/en/products-and-services/product/hardware-and-systems/electric-power/the-honeywell-turbogenerator

[10] Military+Aerospace Electronics. Collins Aerospace announces completion of preliminary design of hybrid flight demonstrator. 8 Aug. 2022. Available from: https://www.militaryaerospace.com/home/article/14280895/collins-aerospace-announces-completion-of-preliminary-design-of-hybrid-flight-demonstrator

[11] Lilium. Introducing the first electric vertical take-off and landing jet. Available from: https://lilium.com/technology/

[12] Ganev ED. High-performance electric drives for aerospace more electric architectures Part I—Electric machines. In: Proceedings of the IEEE Power Engineering Society General Meeting, Tampa, FL, USA. 24–28 June 2007

[13] Modified from James L. Felder, NASA Glenn Research Center, "NASA Hybrid Electric Propulsion Systems Structures," presentation to the committee on September 1, 2015

[14] McFarlane D, Bussmann S. Holonic manufacturing control: Rationales, developments and open issues. In: Deen SM, editor. Agent-Based Manufacturing, Chapter 13. Berlin: Springer-Verlag; 2003. pp. 303-326

[15] A network approach to define modularity of components in complex products by Manuel E. Sosa1 Technology

and Operations Management Area, INSEAD, 77305 Fontainebleau, France, Steven D. Eppinger Sloan School of Management, Massachusetts Institute of Technology, Cambridge, Massachusetts 02139, Craig M. Rowles Pratt & Whitney Aircraft, East Hartford, Connecticut 06108

[16] Implications of modularity on product design for the lifecycle. Proceedings of the 1996 ASME Design Engineering Technical Conferences and Computers in Engineering Conference August 18–22, 1996, Irvine, California. Patrick J. Newcomb, Bert Bras, David W. Rosen Systems. p. 23

[17] James D, Sinha K, Weck OL. Technology Insertion in Turbofan Engine and assessment of Architectural Complexity. 2011

[18] Kaushik S, de Olivier W. A network-based structural complexity metric for engineered complex systems. SysCon 2013 - 7th Annual IEEE International Systems Conference, Proceedings. 426-430. 2013. DOI: 10.1109/SysCon.2013.6549917

[19] Kaushik S. Structural complexity and its implications for design of cyber-physical systems. 2014;**1**:18

[20] Kaushik S. Structural complexity and its implications for design of cyber-physical systems, Chapter 6.2 Distribution of Structural Complexity. 2014. p. 216

Chapter 5

# Principles of Diagnosing: The Technical Condition of the Bearings of the Gas Turbine Engine Supports Using Rhythmogram and Scatterogram

*Dmitry Balakin, Vitaly Shtykov, Alexey Zubko, Shalimova Elena Vladimirovna and Zayed Saleh Salem Ali*

## Abstract

The possibility of using a rhythmogram and a scatterogram for bearings of diagnosing a gas turbine engine and its components is discussed. Rhythmogram and scatterogram evaluate the quasi-periodicity of the technical system of a gas turbine engine. Rhythmogram and scatterogram were obtained using the method developed by us for processing quasi-periodic pulse signals. The method is based on the principles of the theory of optimal filtering, the theory of wavelet transform, and the Hermite transform. The wavelet transform is considered as a cross-correlation function. The Gauss-Hermite functions are used as the basis for wavelet analysis. The effectiveness of the diagnostic method is demonstrated by the example of the operation of the bearing supports of a gas turbine engine and the engine as a whole.

**Keywords:** gas turbine engine, rotor bearings, hermit transform, wavelet transform, optimal filtering, rhythmogram, scatterogram

## 1. Introduction

Most machines and mechanisms as part of their design have rotating elements that transfer the load to the stator through bearings. Structurally, they can be of various types, and the possibility of functioning of the entire device depends entirely on their performance.

Even higher requirements are placed on bearings that are used in the rotor bearings of aircraft gas turbine engines. With a small mass, they must transmit large variable loads and work with high speeds of movement of the contacting parts of their design.

When designing rotor systems of gas turbine engines (GTEs), many years of experience in the development of such structures are taken into account, and computational simulations and their verification are carried out based on the results of various integrated tests. Such methods of calculation and full-scale experiment take into account and repeatedly check the influence of a large number of external and internal factors that can drastically reduce the duration of uptime of engines.

At the same time, despite the progress made in the field of computer design, simulation, and testing, it will never be possible to take into account the entire set of product states due to the manifestation of all possible combinations of values of the influence of various factors that may arise during long-term operation of engines.

Given the importance of ensuring long-term trouble-free operation of the bearings of the rotor systems of GTE, the most promising trend can be considered as the provision of such a design with a developed monitoring and diagnostic system. It should provide a permanent opportunity to determine the technical condition of the bearings in all operating modes of the engine and issue warnings to the crew in cases of rapidly developing damage to their parts.

In view of the importance and complexity of this issue, we have set a goal to provide the possibility of diagnosing the inter-rotor bearing of the GTE turbine support by developing nontraditional diagnostic methods.

Modern methods and algorithms are required not only to diagnose existing defects but also to detect changes in the state of the system before an emergency occurs. In this direction, it is worth noting the works of domestic scientists: [1].

For technical diagnostics of nonstationary systems, a wavelet transform is used, which makes it possible to adapt to the local properties of the signal, which increases the efficiency of diagnostics. This is evidenced by the works of scientists such as [2–4]. Wavelet the transform gives a time-frequency representation of the signal, thereby avoiding many of the problems that arise when using the Fourier transform. The impulses of the system under study can have an arbitrary shape; therefore, to describe them, it is necessary to use orthogonal functions. The most popular orthogonal wavelets are the Haar and Daubechies wavelets [5]. They have a number of limitations and disadvantages. Haar wavelets poorly describe smooth functions, and Daubechies functions have an asymmetric form, which narrows the area of their practical use. Thus, with the help of classical wavelets, it is not possible to take into account all the features of the pulse shape, which affects the quality of diagnostics. In this regard, when processing real signals, it is necessary to construct a "mother" function based on a discrete record of the signal, and this is absent in the classical interpretation of the wavelet transform.

This goal can be achieved by solving the following tasks:

- development of a diagnostic method based on new principles for obtaining and analyzing diagnostic information;
- obtaining diagnostic signs of an early manifestation of a developing malfunction and developing a diagnostic model of an inter-rotor bearing.

## 2. Theoretical foundations of the processing method

The most popular diagnostic methods bearings can be divided into two broad categories: spectral methods and time methods.

In the general case, the analysis algorithm based on the spectral method is as follows: using the properties of the Fourier transform, the signal spectrum is calculated:

$$F(\omega) = S_0(\omega) \sum_{n=0}^{N} \exp(-j\omega\tau_n), \tag{1}$$

where $S_0(\omega) = \int_{-\infty}^{\infty} S(t) \exp(-j\omega t)dt$_bearing signal spectral density ($S(t)$). The calculation $F(\omega)$ is based on statistical properties $\tau_n$ . The parameter $\tau_n$ is a random value that characterizes the quasi-periodicity of the process under study, and the index $n$ shows the pulse number. As a rule, when analyzing the bearing signals by the method (1), it is assumed that process is stationary, so its properties, determined in short time intervals, do not change from the interval to the interval. Thus, $\tau_n = nT$ which is the period of the impulses. Using spectral methods, we can confidently detect defects that are characterized by the occurrence of a periodic component. This is the wear of the rolling bearings, the weakening of mechanical joints in the engine, and much more. Spectral methods include methods based on the analysis of the broadband energy spectrum [6], the frequency component [7], the spectrum of the envelope signal [8], and the window conversion of Fourier [9].

Another direction in technical diagnosis is temporary analysis. Due to this, it can find some characteristics that are difficult to identify using spectral analysis. For example, according to the parameters of a temporary signal, you can detect the number of shock pulses, which in the spectrum may not be distinguishable due to the presence of interference and noise [10].

The disadvantages of classical methods of spectral and temporary processing include that they give only the general condition of the system under study. In addition, classical methods significantly reduce their effectiveness in the analysis of nonstationary processes, which complicates the assessment of dynamic parameters (quasi-periodicity), which means they cannot effectively give a forecast about how the system will behave in the future.

A method is proposed that occupies an intermediate position between spectral and temporary methods and is based:

1. On the main properties of the transformation of the Hermit;

2. On the main properties of the wavelet transformation;

3. On the classical theory of optimal filtration.

Thanks to the use of the basic properties of the wavelet transformation (scaling and localization), the method allows us to adapt to the local features of the bearing signal. As basic functions, the method uses the functions of the Gauss-Hermite (FGH), which are defined in the transformation of the Hermit [11]. The undoubted advantage of the FGH is that they are orthogonal in both the temporary and frequency area, which allows them to adapt them not only to the form of the signal in time, but also to its frequency spectrum. It is also worth noting that the method works throughout the sample of the signal under study, which increases its effectiveness but affects the calculation speed. The stability of the method for noise increases due to the fact that it is based on the principles of optimal filtration. At the output of the method, we have a cross-correlation function, according to which the dynamics of the pulse is evaluated (assessment of quasi-periodicity or $\tau_n$) by measuring the distance between the peaks. For clarity and information content, this procedure can be replaced with an assessment of the rhythmogram of the process under study. The rhythmogram is an apparatus for evaluating quasi-revisionism adopted in the medical field. The rhythmogram reflects the variability of the heart rhythm. Thus, the use of a rhythmogram to assess the state of the bearing is a new field in the diagnosis of technical systems.

### 2.1 Mathematical description of the method

The mathematical procedure for processing the signal of the bearing can be described by the following scheme:

The input digital signal $S_{in}(t_k)$ (signal after sampling and quantization), obtained from the vibration sensors, changes with a reference signal $\tilde{S}_{ptrn}(t_k, m)$. A reference signal is a fragment of a record $S_{in}(t_k)$, the dynamics of which we want to trace. A reference signal can be interpreted as a defect or a useful signal, the choice depends on the objectives of the study. A reference signal can be built in the following expression:

$$\tilde{S}_{ptrn}(t_k, m) = \frac{1}{\sqrt{2^m}} \sum_{q=0}^{Q} W(q, q_{\text{cf}}) A_q(m) \Psi_q(t_k, m) \tag{2}$$

The main computing element (2) is the calculation, which is based on the transformation of the Hermit or FGH and the discrete wavelet transformation:

$$A_q(m) = \left(q! 2^q \sqrt{\pi}\right)^{-1/2} \sum_{k=-K}^{K} S_{ptrn}(t_k) \Psi_q(t_k, m), \tag{3}$$

where $S_{ptrn}(t_k)$ is the vector of samples of the selected fragment (or the standard is a fragment from $S_{in}(t_k)$), $t_k = \Delta t \cdot k$, $\Delta t$ is the sampling step, $K = N_{ptrn}/2$, $N_{ptrn}$ is the duration of the selected fragment in samples, $k = -K,. -1,0,1,..K$, borders on $\Delta t = 3/K$, time by level $\pm 3$, $\Psi_q(t_k, m)$—FGH $q$—FGH order, $m$ is the scale parameter (similar to the scale parameter in the wavelet transform). In fact, (3) is the decomposition of a fragment $S_{ptrn}(t_k)$ in space $\Psi_q(t_k, m)$. Gauss-Hermite functions have the following mathematical notation:

$$\Psi_q(t_k, m) = H_q\left(\frac{t_k}{2^m}\right) \exp\left(-0.5\left(\frac{t_k}{2^m}\right)^2\right), \tag{4}$$

where $H_q\left(\frac{t_k}{2^m}\right)$—Hermite $q$—order polynomial.

Then, (2) is the inverse Hermite transform, and the reproduction accuracy $S_{ptrn}(t_k)$ is determined by the expression:

$$Er(m, q) = \frac{\sum_{k=-K}^{K} \left(\frac{1}{\sqrt{2^m}} \sum_{q=0}^{Q} W(q, q_{\text{cf}}) A_q \Psi_q(t_k, m) - S_{ptrn}(t_k)\right)^2}{\sum_{k=-K}^{K} S_{ptrn}(t_k)^2} 100\%, \tag{5}$$

where $W(q, q_{\text{cf}})$ is the smoothing filter in FGH space. When the series is truncated in FGH space, the reference signal has oscillations, and this effect is analogous to the

"Gibbs phenomenon" in harmonic analysis. To weaken it, you can use low-pass filtering in the FGH space, based on well-known approximations of the frequency characteristics: Gauss, Butterworth, and Bessel.

The error $Er(m, q)$ can be set by the researcher and is determined by the maximum FGH order—$Q$ .

The key feature of the signal according to (2) is that it is built from a real discrete record of the signal, so we can take into account any local features of the process under study.

In accordance with the scheme of **Figure 1**, the next operation is the integration or calculation of the correlation integral. The correlation integral is defined in the theory of optimal filtering and searches for similar fragments $\tilde{S}_{ptrn}(t_k, m)$ is the process under study $S_{in}(t_k)$. In this paper, the correlation integral is built on the basis of the wavelet transform:

$$R_{out}(t_n, m) = \sum_{k=0}^{N} S_{in}(t_k)\tilde{S}_{ptrn}(m, t_k - t_n), \quad (6)$$

where $N$ is the maximum number of samples in the signal $S_{in}(t_k)$. As you can see, with a large number of readings, calculation (6) is computationally labor-intensive. Therefore, it makes sense to move from spatial integration in the time domain to multiplication in the frequency domain or matched filtering. Then the scheme of **Figure 1** can be transformed into the following form (**Figure 2**):

The signal goes directly to the matched filtering (*MF*) block. With matched filtering, an increase in the detection rate is associated with the following circumstances:

1. the spectral image of the Gauss-Hermite functions is known;

2. the orthogonal basis of the Gauss-Hermite functions in space-time corresponds to the orthogonal basis of the Gauss-Hermite functions in the frequency domain;

3. when calculating in the spectral region, you can use the fast Fourier transform procedure.

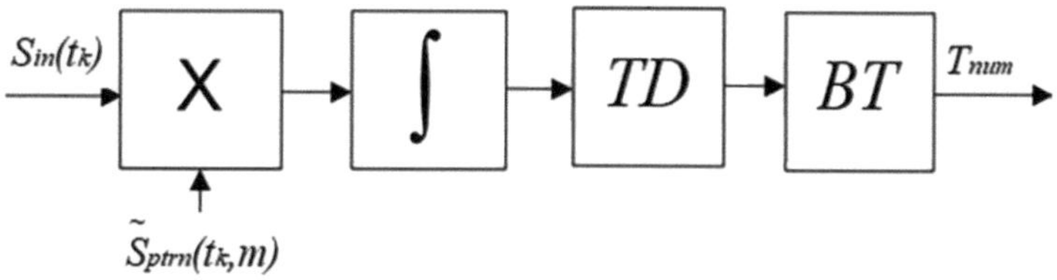


**Figure 1.**
*Block diagram of the processing method in the temporary area.*

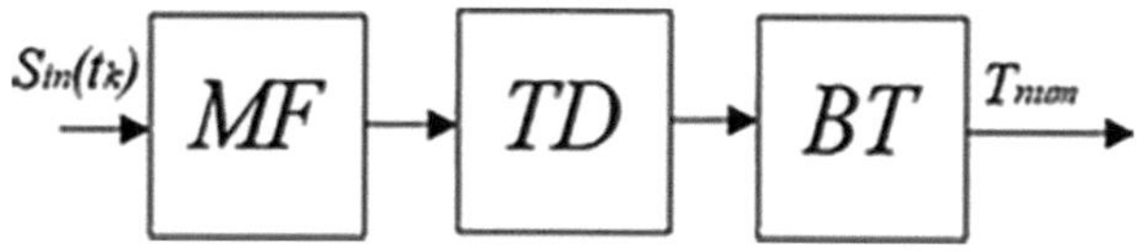


**Figure 2.**
*Flowchart of the processing method in the time-frequency domain.*

It was obtained [12] the transmission coefficient of the filter that selects a fragment of the signal, consisting of a set of features:

$$\dot{K}(f_n, m) = \sum_{q=0}^{Q} j^q \sqrt{\frac{\sqrt{\pi}}{q!2^{q-1}}} \exp\left(-0,\ 5(2^m)^2 {f_n}^2\right) H_q(2^m f_n), \tag{7}$$

where $f_n = n/N\Delta t, n = 0,1, \dots N-1$.

The signal after matched filtering the cross-correlation function enters the threshold device ($TD$), where, based on a priori statistical information, peaks exceeding a certain threshold value are selected. The median value can be used as a threshold. After threshold processing, the resulting one enters the block of transformation ($BT$), at the output of which we have the rhythmogram of the process ($T_{num}$).

Due to the fact that the principles of the wavelet transformation are incorporated in the processing method, the rhythmogram can be refined by varying the scale parameter or the filter band (7):

$$F(t_k, m) = \sqrt{E_s} - \left| \frac{1}{2\pi\sqrt{a}} \sum_{n=0}^{N-1} \dot{S}_{in}(f_n) \dot{K}_q(f_n,\ m) \exp\left(\frac{j2\pi f_n t_k}{N}\right) \right|, \tag{8}$$

As a result, a functional or a surface is formed, on which there are many extrema (minimums) corresponding to the number of reference signals in the process under study. The processing process is reduced to solving a multi-extremal problem. This problem is solved using the steepest descent method, where a priori information about the rhythm of the system under study can serve as starting points. However, in real problems, such information is not always available. Therefore, as a starting point for the steepest descent algorithm, it is advisable to use the coordinates of the maxima of the cross-correlation function determined at an early stage without changing the scale parameter. As a result of such processing, we obtain refined information about the location of the reference signal and, consequently, a more accurate process dynamics or rhythmogram.

## 2.2 Signal processing without a priori information

The records of three signals ($S_1, S_2, S_3$) are shown in **Figure 3**. The sample size is 30,000 readings. In accordance with the considered theory, it is necessary to isolate the standard from the signal under study. The standard is a fragment of the recording, the dynamics (rhythm) of which we want to trace throughout the entire time of the study. The standard reflects the distinctive features of the waveform, which can mean both the correct operation of the device, and, conversely, a faulty one. The choice of standard should be carried out taking into account the opinion of a specialist or an expert in the field of diagnostics. Since we do not have a priori information on the characteristic features of the signal, we will conduct a preliminary spectral analysis of the signals to select the standard.

Take the value of the sampling frequency equal to 1. Then, for a sample of 30,000 samples, the frequency step in the spectral region will be 1/30,000 rel. units. The recording spectra are shown in **Figure 4**. The figure shows that the maximum frequency components in the signal spectra are concentrated in the range of approximately 0.015–0.02 rel. units, which corresponds to approximately 50–70 samples in the signal record.

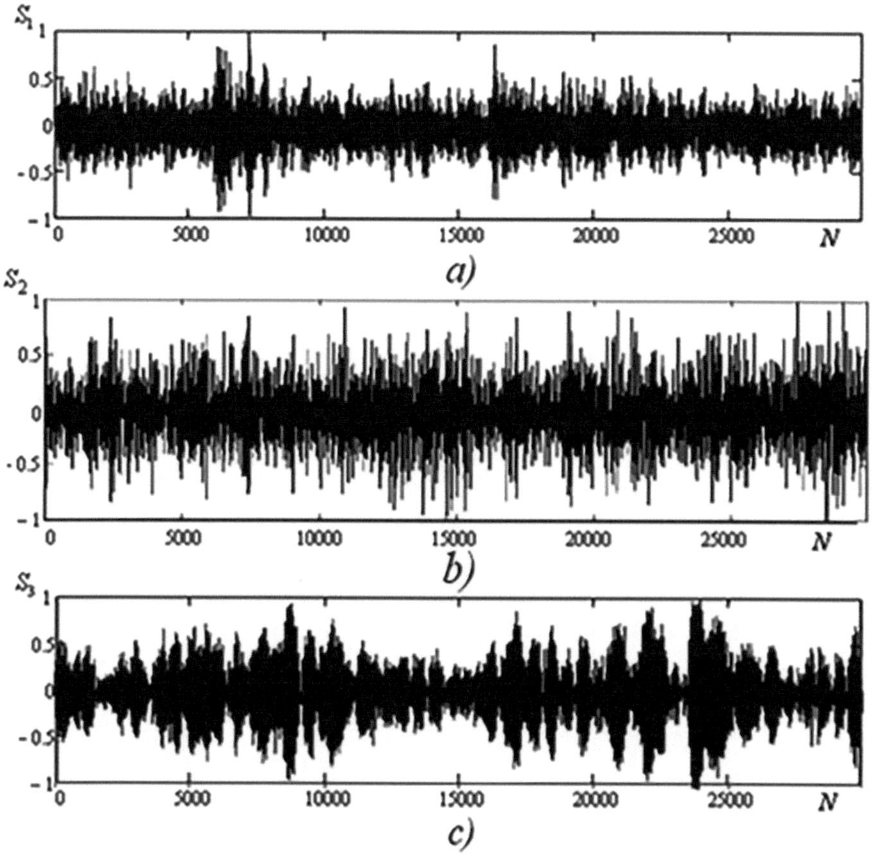


**Figure 3.**
*Recordings of vibration signal of bearings: N—count number.*

The ideal mechanism functions cyclically, which is determined by the characteristic spectral component. In the discrete spectrum of the signal of such a mechanism, the reciprocal of the period has a maximum value. In real records, the spectrum is blurred (**Figure 4**), but its maximum can be identified with the average pulse repetition period.

After analyzing the spectra, it can be established that the signal $S_2$ is a faulty operation of the bearing, since many spectral peaks are observed in the spectrum. Bearing signals $S_1$ and $S_2$ are of the greatest interest due to the fact that it is impossible to say for sure from the spectral pattern whether they are in good order or not. Let us choose a signal fragment from the record as a reference, since its main spectral components are present both in the signal spectrum and in the signal spectrum.

After analyzing the record, a fragment (from 6020 to 6080 samples) was selected, shown in **Figure 5**, since similar waveforms occur throughout the entire signal record.

In **Figure 5**, characteristic fragments with a duration of approximately 50–70 samples are clearly visible. The next stage of processing is the construction of a support function (SF) based on FGH. In essence, SF is a mirror image impulse response of a complex quasi-matched filter (7). As a result of multiplying the complex conjugate transfer coefficient of the quasi-matched filter and the spectrum of the signal under study, after the inverse Fourier transform, we obtain the cross-correlation function, fixing its maxima, we can build a rhythmogram.

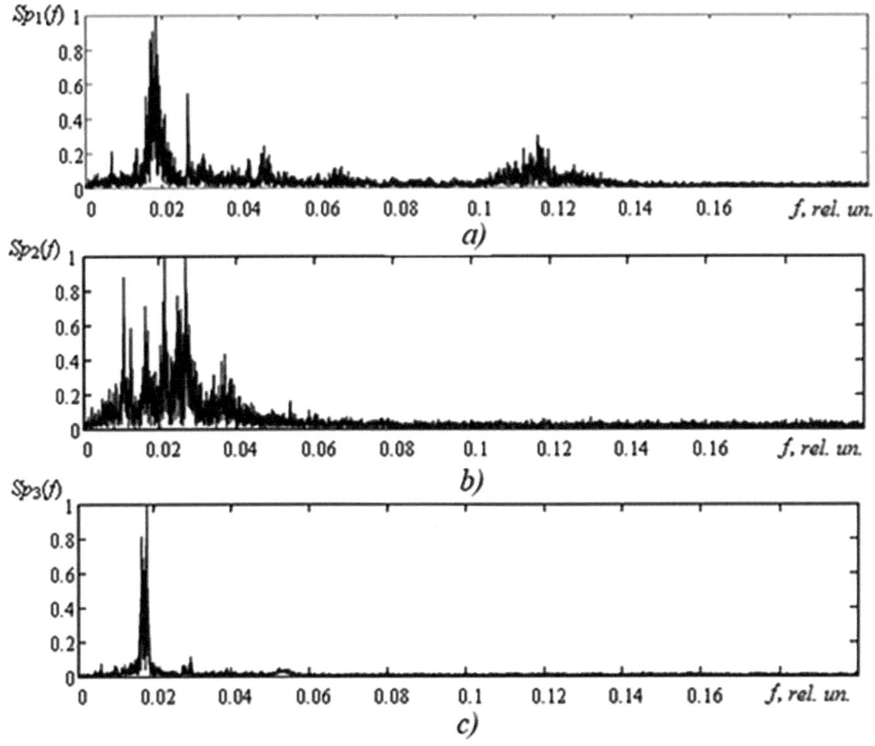


**Figure 4.**
*Spectra of the studied signals.*

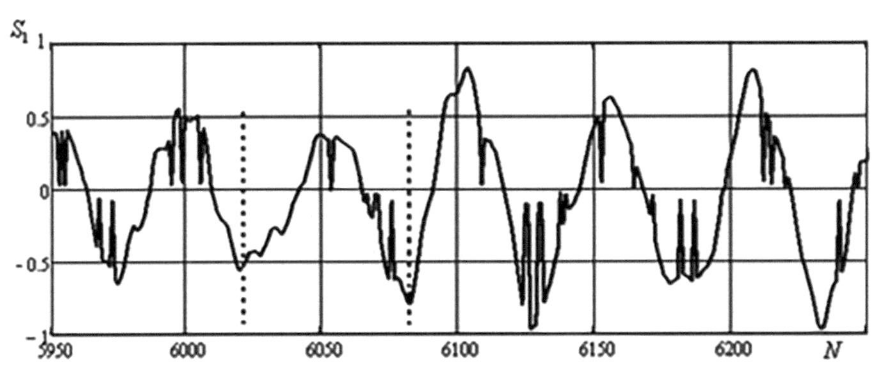


**Figure 5.**
*Fragment of the record $S_1$.*

The position of the extrema of cross-correlation function can be refined by varying the scale parameter when constructing the filter (8). Thus, the problem of diagnosing a system is reduced to solving a multi-extremal problem, since the cross-correlation function is a complex surface with many local maxima. From a methodological point of view, it is more convenient to look for local minima. This transition can be made using the Cauchy-Bunyakovsky inequality (**Figure 6**).

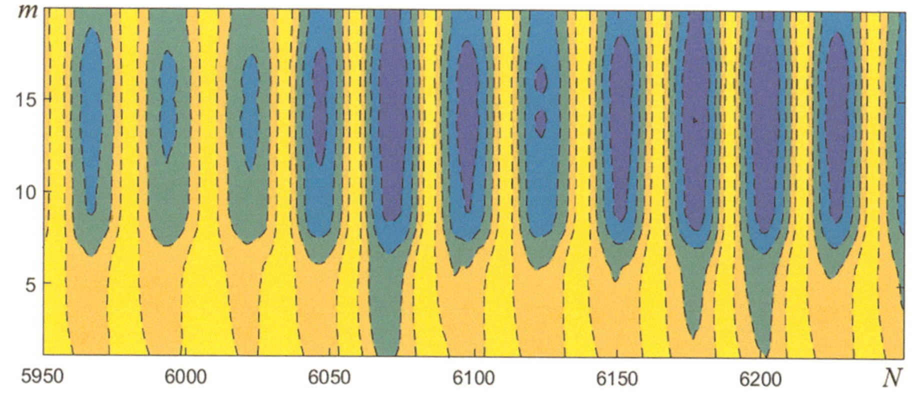


**Figure 6.**
*Topographic picture of the cross-correlation function of a signal $S_1$ fragment with a variation of the scale parameter.*

Such a procedure (**Figure 6**) resembles the process of scale variation in the wavelet transform. The key difference between the two processing approaches is that at the output of the method under consideration, we have the cross-correlation function that allows us to estimate the degree of similarity of the SF and the reference, and with wavelet processing, the spectral distribution. The position of the minima of the obtained surface can be found, for example, by the steepest descent method.

As a result of such processing, we have rhythmograms and scatterograms.

To construct the rhythmogram, the cross-correlation function was subjected to threshold processing. The processing results are shown in **Figure 7**.

When constructing the diagram, the minimum positive value from the set of maxima of the cross-correlation function signal $S_1$ was taken as the threshold.

Rhythmograms look like a random process with an average value. It is approximately equal to 50 samples and gives an estimate of the average period of the processed signals.

Rhythmogram surges indicate deceleration (upsurges) or acceleration (downsurges) of the mechanism. Upsurges commensurate with the average value of the rhythmogram indicate that the algorithm skipped a cycle due to the fact that the maximum cross-correlation function does not exceed the set threshold. There are few such outliers in **Figure 7a** and **c**, while there are quite a lot of them in **Figure 7b**. In the presence of noise, downward spikes may appear, commensurate with the average value of the rhythmogram, due to the appearance of false maxima. There are no such outliers in **Figure 7**.

Preliminary visual analysis suggests that the bearings, the signals of which are shown in **Figure 1a** and **c**, are in good condition, and the bearing with the signal $S_2$ has a defect. Let us confirm these preliminary considerations with quantitative estimates.

The rhythmogram can be considered as a discrete signal that can be processed by one of the traditional methods. You can, for example, get the spectrum of the rhythmogram. **Figure 8** shows the smoothed spectra of rhythmograms after low-frequency filtering. As can be seen, the signal $S_2$ spectrum stands out significantly compared to the signal spectra $S_1$ and $S_3$. For a quantitative assessment, we will perform a statistical analysis of rhythmograms. We calculate the mean, standard deviation, mode and median, as well as the minimum and maximum values. The calculated parameters are presented in **Table 1.**

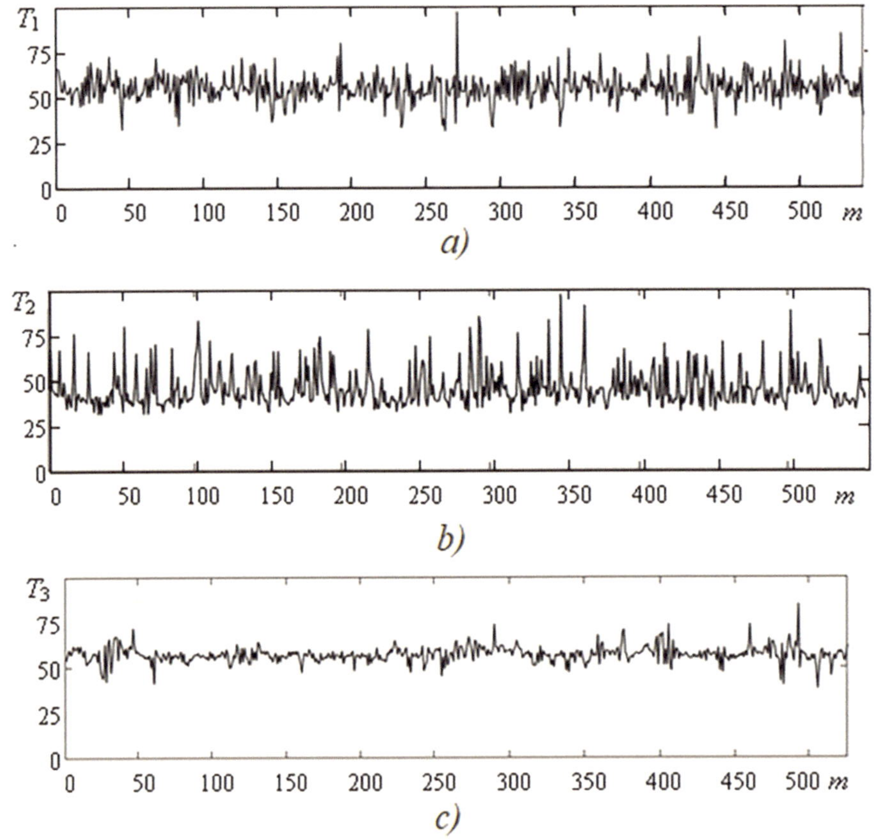


**Figure 7.**
*Rhythmograms of signals, m is the number of the cross-correlation function maximum, $T_1$,$T_2$,$T_3$—the duration of the intervals between the maxima of the cross-correlation function.*

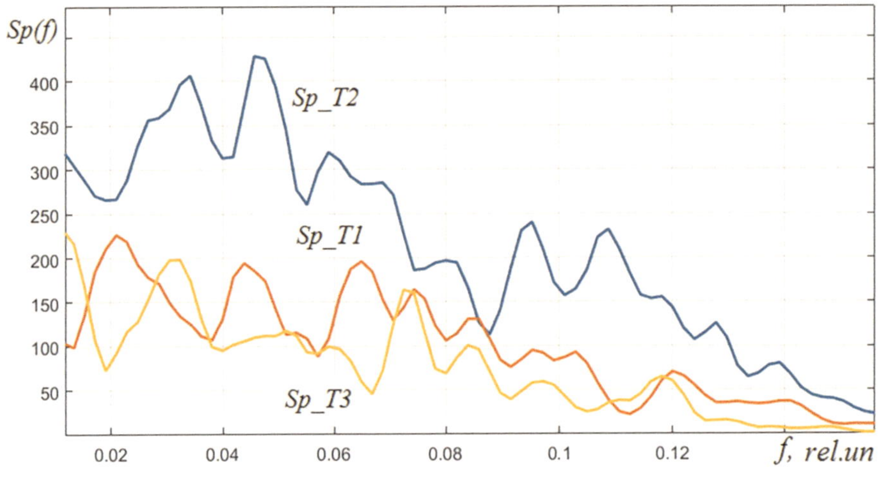


**Figure 8.**
*Spectra of rhythmograms: Sp_T1—spectrum of $T_1$ rhythmogram, Sp_T2— spectrum of $T_2$ rhythmogram, Sp_T3—spectrum of $T_3$ rhythmogram.*

| Parameter | signal | | |
|---|---|---|---|
| | $S_1$ | $S_2$ | $S_3$ |
| Mean | 55.2 | 47.5 | 55.9 |
| Standard deviation | 8.55 | 11.5 | 6.75 |
| Median | 54 | 44 | 56 |
| Mode | 52 | 43 | 56 |
| Minimum value | 32 | 31 | 32 |
| Maximum value | 101 | 109 | 88 |

**Table 1.**
*Statistical parameters of rhythmograms.*

In medical practice, the diagnosis of pathology is based on the value of standard deviation. The standard deviation values given in **Table 1** differ by 10%. The signal $S_2$ has the highest standard deviation value. The median and mode of recording signals $S_1$ and $S_3$ have close numerical values of the statistical parameters, which are close to the average. This indicates that the distribution of outliers relative to the mean value is close to symmetrical. For the signal $S_2$, downward surges (jerks and bumps) predominate. Next, consider the scatterograms of GTE bearing signals.

Scatterogram is a geometric method. In practice, the scatterogram has the shape of an ellipse stretched along the bisector **Figure 9** [12].

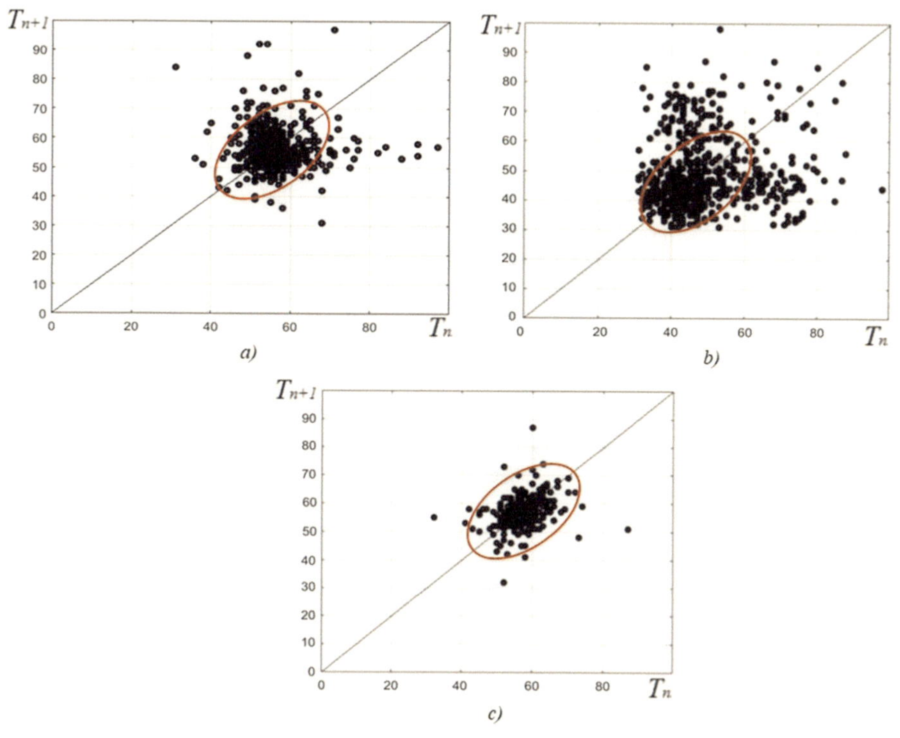


**Figure 9.**
*Scatterograms of signals, Tn Tn + 1—duration of the previous and subsequent intervals between the maximum cross-correlation.*

According to the scatterogram, one can judge the quasi-periodicity of the signal under study. The more clustered the points are, the less the quasi-periodicity.

The shift of the points to the right along the coordinate axis reflects a decrease in the rhythm, while the shift to the left reflects an increase. If the points are far from the whole population, then this may indicate a defect.

Based on the totality of statistical estimates of rhythmograms and the scatter of points in scatterograms, we can assume that the bearings $S_1$ and $S_3$ are both in good condition, and $S_2$—with a high degree of probability—are faulty.

### 2.3 Signal processing with a priori information

Next, consider the processing of vibration signals received from the stand SP-180 M (**Figure 10**). This bench allows you to test all sizes and types of modern and advanced bearings used in aircraft gas turbine engines. The rotational speed at the maximum mode of the inter-rotor and inter-shaft bearings installed between the shafts of the rotors of high and low pressure of the engine, rotating in the parallel direction, is 3000 rpm. Therefore, on the stand, the operation of the bearings is modeled at speeds from 0 to 3000 rpm [13].

Stand SP-180 M allows us to place sensors directly on the outer or inner rings of the bearing [14]. As a rule, it is installed by gluing on a non-rotating outer ring. The results obtained make it possible to judge the condition of the bearings by the generated vibrations measured by the vibration sensors (**Figure 11**).

GTE tests are also being carried out on the high-pressure rotor of which was prepared with additional vibration sensors on the high-pressure compressor support. The results obtained are compared with information from vibration sensors located on the outer casing of the engine (**Figure 12**).

**Figure 10.**
*External view of the stand SP-180 M.*

**Figure 11.**
*Appearance of the installed bearing for vibration measurements by the built-in sensors of the bench.*

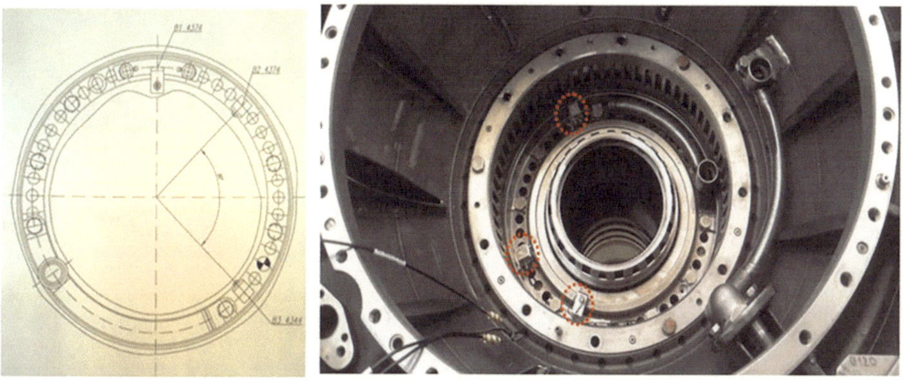

**Figure 12.**
*Scheme and appearance of vibration sensors installed inside the engine, on the support of the GTE high-pressure compressor.*

**Figure 13** shows the fragments of vibration signals of the bearings received from SP-180 M. One bearing is good and the other two are defective. We will process the vibration signals using the above method.

From a priori information about the signals $S_1(t)$, $S_2(t)$, and $S_3(t)$, it is known that the rotation frequency of the separator is 23 Hz, the frequency of rotation of the rolling elements is 329 Hz, the frequency of rolling of the rolling elements along the outer ring is 529 Hz, and the frequency of rolling of the rolling elements along the inner ring is 612 Hz. In this regard, before selecting the standard, we will perform a preliminary filtering according to the scheme, which is shown in **Figure 14**.

In accordance with the scheme of **Figure 14**, the signal $S_1(t)$ of a healthy bearing is prefiltered, and each of the filters is a bandpass filter with the center frequency of the

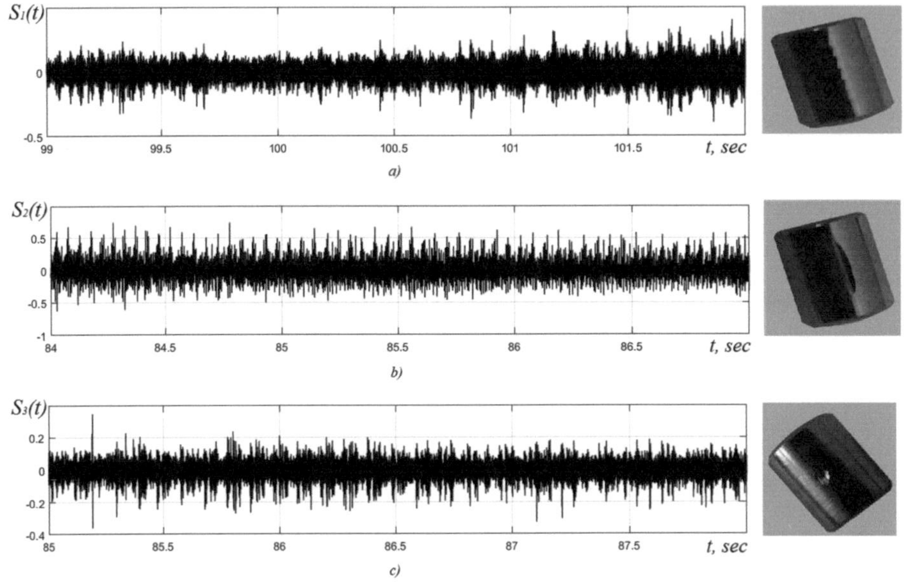


**Figure 13.**
*Recordings of vibration signal of bearings: a) correct operation, b) damage—medium risk, and c) damage—point large and small.*

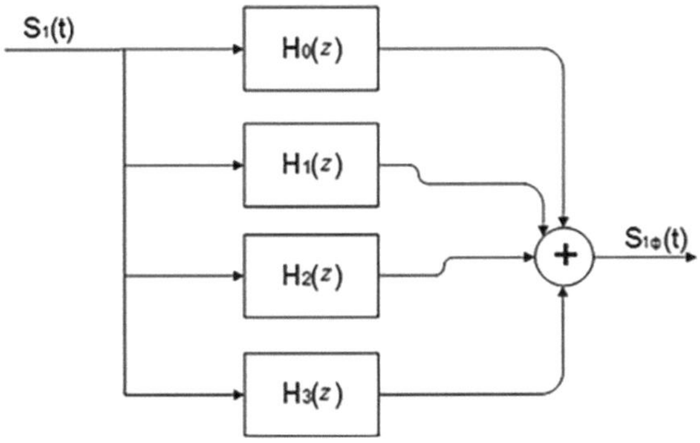


**Figure 14.**
*Prefiltration scheme.*

passband corresponding to the frequency component from the prior knowledge. It is also worth noting that the filtering procedure is performed separately, and the signal is summed at the output of each of the filters. The transfer coefficients of the bandpass filters are shown in **Figure 15**.

Next, a filter was designed according to the formula (7), the frequency response of which is shown in **Figure 16**. The frequency response peaks at the given a priori frequencies.

As a result, we have the following rhythmograms and scatterograms (**Figures 17** and **18**) and statistical evaluation **Table 2.**

**Table 2** shows that the standard deviation value of the rhythmogram of a serviceable bearing is more than one and a half times lower than the standard deviation value of bearings with defects, which, in turn, can be an additional diagnostic feature.

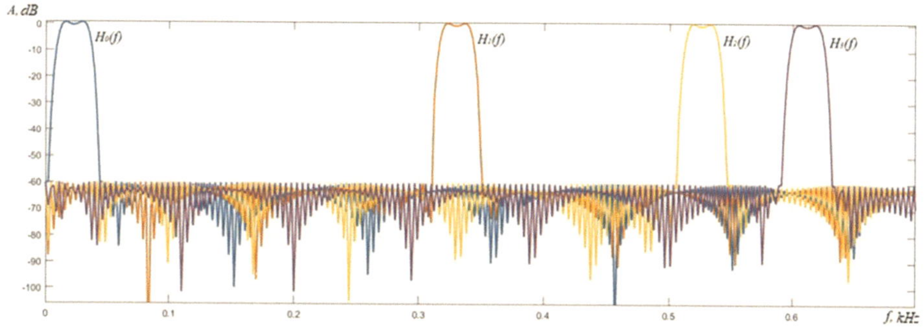


**Figure 15.**
*Frequency response of bandpass filters.*

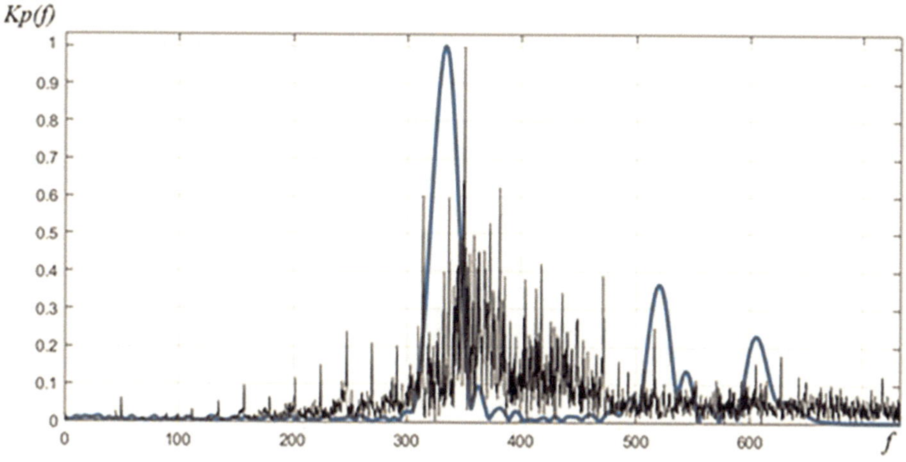


**Figure 16.**
*Quasi-matched filter and Fourier spectrum of a good bearing.*

The median of the signal $S_1(t)$ record has a slight deviation from the numerical value of the mathematical expectation, while the deviation of the median from the average value for signals $S_2(t)$ and $S_3(t)$ above. An approximately similar picture is observed in the analysis of modes. This indicates that the distribution of outliers relative to the mean value is close to symmetric for $S_1(t)$ . For signals $S_2(t)$ and $S_3(t)$ emissions down (jerks and bumps) prevail [15].

## 3. Conclusion

Rhythmogram and scatterogram allow you to explore the system in various modes, both standard and special. Therefore, it is able to provide additional diagnostic information about the state of the dynamic system. Thus, it can be hoped that such a diagnostic method may be of practical interest for assessing the characteristic dynamic features of the functioning of various mechanisms, devices, and apparatuses. Of course, the rhythmogram and the scatterogram will have a different look if another fragment of the record is chosen as a reference due to the change in SF. The issue of choosing a standard requires additional research with the involvement of diagnostic

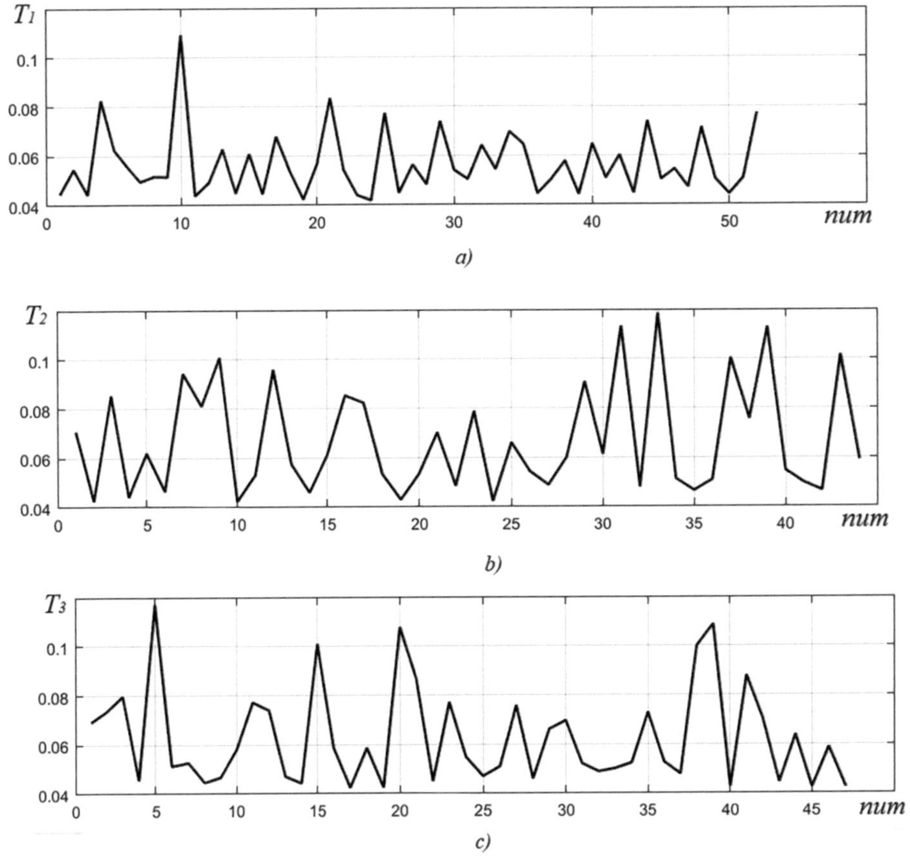


**Figure 17.**
*Rhythmograms of signals num is the number of the maximum cross-correlation, $T_1$,$T_2$,$T_3$—the duration of the intervals between the maxima of the cross-correlation.*

| Parameter | Signal | | |
|---|---|---|---|
| | $S_1(t)$ | $S_2(t)$ | $S_3(t)$ |
| Mean | 56.7 | 67.1 | 62.8 |
| Standard deviation | 13.3 | 22.3 | 20.1 |
| Median | 54 | 59.4 | 54.9 |
| Mode | 50.4 | 46.5 | 42.5 |
| Minimum value | 41.9 | 42.4 | 42.5 |
| Maximum value | 109.2 | 118.5 | 117.1 |

**Table 2.**
*Statistical parameters of rhythmograms.*

specialists in each specific case. The development of databases of various SF will allow for express diagnostics of devices. A feature of the method proposed and developed by us for processing quasi-periodic pulse signals using the Gauss-Hermite function is also

*DOI: http://dx.doi.org/10.5772/intechopen.108400*

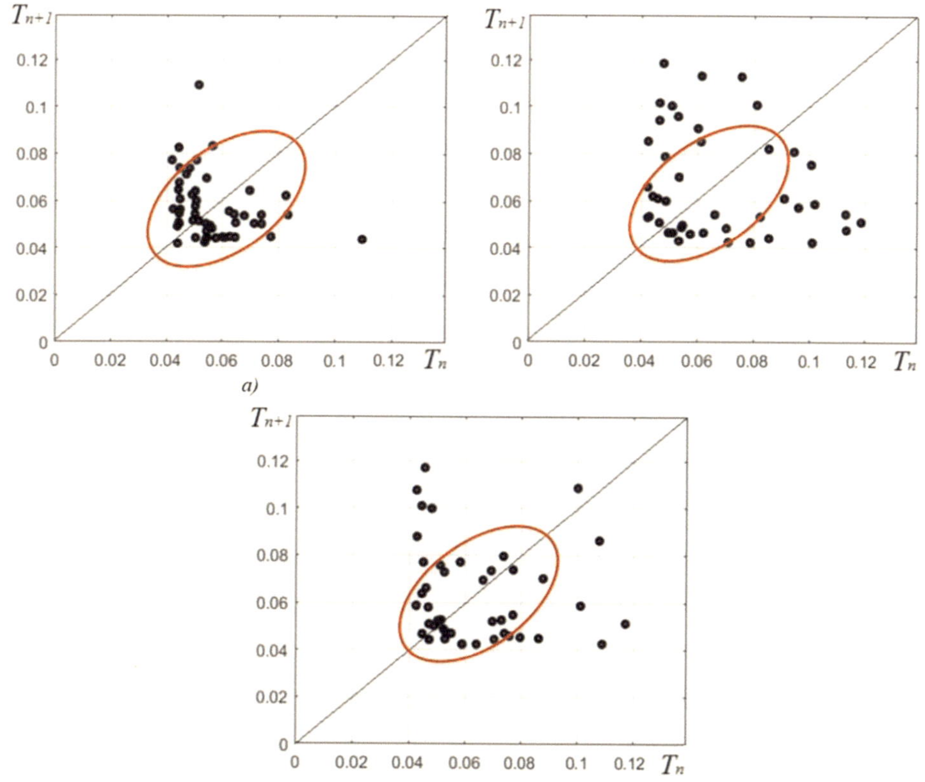


**Figure 18.**
*Scatterograms of signals, Tn Tn + 1—duration of the previous and subsequent intervals between the maxima of the cross-correlation.*

the possibility of varying the SF scale along the time axis. The scale variation makes it possible to detect not only violations of the system periodicity, but also a change in the shape of the signal pulses, which, in turn, provides additional means for detecting and predicting the development of a malfunction.

## Author details

Dmitry Balakin[1*], Vitaly Shtykov[1], Alexey Zubko[2], Shalimova Elena Vladimirovna[1] and Zayed Saleh Salem Ali[1]

1 National Research University "Moscow Power Engineering Institute", Moscow, Russia

2 Moscow Aviation Institute (National Research University), Public Joint Stock Company «UEC-Ufa Engine Industrial Association» branch «OKB after A. Lyulki», Moscow, Russia

*Address all correspondence to: dabalakin@yandex.ru

## References

[1] Barkov AV, Barkova NA, Mitchell JS. Condition assessment and life prediction of rolling element bearings. Part 1. Sound and Vibration. 1995;**6**:10-17

[2] Adewusi S, Al-Bedoor BO. Wavelet analysis of vibration signals of an overhang rotor with a propagating transverse crack. Journal of Sound and Vibration. 2001;**246**(5):777-793. DOI: 10.1006/jsvi.2000.3611

[3] Chancey VC, Flowers GT. Identification of transient vibration characteristics using absolute harmonic wavelet coefficients. Journal of Vibration and Control. 2001;**11**(7):1175-1193. DOI: 10.1177/107754630100700803

[4] Peng Z, Chu F, He Y. Vibration signal analysis and feature extraction based on reassigned wavelet scalogram. Journal of Sound and Vibration. 2002;**253**(5): 1087-1100

[5] Nazolin AL. Possibility of detecting defects in the stator of Turbogenerator from the Spectrum of Vibroacoustic signal. Industrial laboratory. Diagnostics of materials. 2017;**83**(4): 49-54 (In Russ.)

[6] Sundukov AE, Sundukov EV. Analysis of the vibration state of quality bearings. Vestnik SSAU. 2006;**2**(10): 85-89 (In Russ.)

[7] Kostyukov VN, Naumenko AP. Handbook of Fundamentals of Vibroacoustic Diagnostics and Monitoring Ring of Machines. OmSTU Publishing House; 2011. p. 360 (In Russ.)

[8] Barkov AV, Barkova NA, Azovtsev AY. Handbook of Monitoring and Diagnostics of Rotary Machines by Vibration. SPbGMTU publishing house; 2000. p. 150 (In Russ.)

[9] Yurkevich VV. Diagnostics of rolling bearings. Stankoistroument. 2015;**1**: 97-99 (In Russ.)

[10] Martens JB. The Hermite transform: A survey. EURASIP Journal on Advances in Signal Processing. 2006;**2006**:026145

[11] Balakin DA. Construction of an orthogonal filter bank based on Hermite transforms for signal processing. Journal of Radio Electronics. 2014;**9**:8 (In Russ.)

[12] Balakin DA, Shtykov VV. Using rhythmograms to diagnose mechanical systems. In: Journal of Physics: Conference Series. IOP Publishing Ltd. Vol. 1399. APITECH; 2019. p. 4. DOI: 10.1088/1742-6596/1399/4/044027

[13] Semenova AS, Zubko AI. Studying technical condition of the interrotor bearing whith the SP-180M vibratory-diagnostic test bench after passing life tests. Vesnik of MAI. 2015;**26**(1):126-138 (In Russ.)

[14] Zubko AI. Complex Method of Vibrodiagnostics of the Technical Condition of Bearing Supports of Gas Turbine Engines [Thesis]. Moscow Aviation Institute: Moscow; 2020

[15] Balakin DA, Zubko AI, Zubko AA, Shtykov VV. Vibration diagnostics of rotor bearings, Hermite transformation, rhythmogram, scatterogram, quasi-periodicity. Vesnik of MAI. 2021;**28**(4): 151-162 (In Russ.)